Frozen Earth

Frozen Earth

The Once and Future Story of Ice Ages

Doug Macdougall

UNIVERSITY OF CALIFORNIA PRESS

Berkeley   Los Angeles   London

University of California Press
Berkeley and Los Angeles, California

University of California Press, Ltd.
London, England

First paperback printing 2006
© 2004 by the Regents of the University of California

Library of Congress Cataloging-in-Publication Data

Macdougall, J. D., 1944–.
    Frozen earth : the once and future story of ice ages / Doug Macdougall.
        p.   cm.
    Includes bibliographical references and index.
    ISBN 0-520-24824-4 (pbk : alk. paper)
        1. Glacial epoch.   2. Paleoclimatology.   3. Global environmental
change.   I. Title.

QE698.M125   2004
551.7'92—dc22

                                        2004008502

Manufactured in the United States of America
15   14   13   12   11   10   09   08   07   06
10   9   8   7   6   5   4   3   2   1

For Grace and Lorn Macdougall,
who always encouraged exploration

CONTENTS

ILLUSTRATIONS

ACKNOWLEDGMENTS

Many thanks go to those who graciously allowed me to use their photographs in this book: Professor Kenneth Hamblin, Brigham Young University; Dr. John Shelton, La Jolla, California; Professor Michael Hambrey, Liverpool John Moores University; Mr. Vasko Milankovitch, North Balwyn, Australia; and Professor John Crowell, University of California Santa Barbara.

My agent, Rick Balkin, worked hard to ensure that this project got off the ground in the first place, and Blake Edgar at U.C. Press provided much input along the way, helping to make the final manuscript more readable, and, I hope, a more interesting book. Guy Tapper produced all of the line drawings in his usual professional manner. Heartfelt thanks to all of you.

CHAPTER ONE

Ice, Ice Ages, and
Our Planet's Climate History

The American author and historical popularizer Will Durant once wrote, "Civilization exists by geological consent, subject to change without notice." That is not a new idea, even if Durant phrased it especially well, but nowadays many historians scoff at the notion of environmental determinism, the possibility that climate or geology may have seriously affected the course of human history. And yet there are still many places on this planet where Durant's observation rings true, especially places with extremes of climate. One such is the arctic regions, particularly Greenland. Ninety-five percent of that island country is covered by ice. Towns and villages cling to the coastline; at their backs loom glaciers a thousand meters thick: gleaming, white, blue, clear, transparent ice. The icecap weighs on the land like a lead brick on a floating plank, pressing it down below the level of the surrounding sea. If the ice were suddenly removed, the waters of the ocean would rush in to take its place. The glaciers seem fixed and static, but in reality they are dynamic, in constant slow movement outward from their thick centers. New snowfall adds to their mass every year, but at the margins they calve off apartment-block-sized chunks of themselves and send flotillas of weirdly shaped icebergs sizzling and crackling and sometimes eerily and silently floating down the fjords to the sea. The icebergs carry pieces of Greenland with them too, sand, pebbles, and boulders gouged and scraped from the land, later to be dropped far out at sea as the ice melts. The Inuit of Greenland have lived with the ice of glaciers for thousands of years. They are truly people of the ice age. Most of the rest of us have been affected by the ice age too, but in less obvious ways.

Permanent icefields—that is, large glaciers—are not common in mainland North America. In the mountainous west, in Alaska and in the Yukon, there are small high-altitude glaciers, but in the overall scheme of things, they are fairly minor features of the landscape. However, as a boy, like many others both in North America and northern Europe, I grew up surrounded by the work of ice. Like most others, I was, at the time, completely unaware of that fact. I am not referring to the ice of a skating rink or of a January puddle. Rather, this was ice just like that of Greenland today, or of Antarctica, ice of vast extent and kilometers thick that blanketed huge swathes of the Northern Hemisphere thousands of years ago. It reached down from centers in Canada and Scandinavia and covered the sites of cities such as Boston, Detroit, and Hamburg. Its legacy is everywhere even today, from the geography of our waterways to the distribution of native peoples in the New World. It ground up solid rock to make the sand of countless beaches and the soil of midwestern farms in the United States. It sculpted rolling hills and long valleys across the landscape. It scraped up soil and rocks as it flowed, and dumped the debris as terminal moraines in places like Cape Cod and Long Island, New York, far from its original home. It even picked up diamonds from still-undiscovered deposits in Canada and transported them to the United States, twenty thousand years before NAFTA was conceived.

The present-day ice sheets of Greenland, and the glaciers in Alaska and arctic Canada, are residual from that once much more extensive ice covering of the Northern Hemisphere. But it was only in the nineteenth century that the existence of those great ice sheets of the past began to be recognized. Although some of our distant ancestors lived cheek by jowl with the gigantic ice caps, the small glaciers that still survived in high mountain regions by the dawn of modern civilization gave few clues to the earlier extent of ice. The massive ice sheets of Greenland and the Antarctic were far from the consciousness of most of the world's population and remained largely unexplored until the late nineteenth and early twentieth centuries. Except for a few small mountain glaciers in Switzerland, there were no glaciers close to the centers of learning that could serve as examples. The story of the ice ages had to be worked out from other, much more tenuous, evidence. Like most other scientific advances, the realization that the Earth has periodically been gripped in ice ages didn't come in a single Eureka! moment. Rather, it developed over a period of time and through the efforts of many naturalists and other close observers of the natural landscape. It came at a time when the science of geology was still young, when the concept that the Earth had an almost inconceivably long history was still controversial, and when the practice of making careful and systematic observations of the natural world was still relatively novel. The ice age had left its marks abundantly on the lands of the Northern Hemisphere. The signs were familiar to farmers and travelers, but for the most part their origins were obscure. It took keen observation, insight, and imagination to recognize in these marks the events that they actually record. And in spite of the fact that by the early part of the nineteenth century many scientists had discarded the notion that nearly all features of the landscape resulted from the biblical Flood, such ideas died hard. Some theologians and others prominent in society thundered "blasphemy" at the idea of an ice age. Even if they didn't have strictly theological objections, when the idea that northern Europe had once been buried beneath a huge glacier was first proposed, many contemporary scientists summarily dismissed it. There was no analog. They could not conceive of such a drastic transformation of the countryside where they now saw only farmland, forests, and rural villages. From the perspective of a single human lifespan, or even on the timescale of a few generations, the Earth appeared to be quite an unchanging place.

In hindsight, it is easy to say that the geological evidence for ice ages was overwhelming and to wonder why such periods in the Earth's past were not recognized earlier. And to be fair, even in the eighteenth century, nearly a hundred years before the term "ice age" was coined, there were already a few bold scientists who had begun to recognize the significance of the evidence. They and others who studied the Earth by careful observation were gradually eroding the influence of theologians who tried to shoehorn virtually every observation of the natural world into a literal biblical framework. Still, widespread debate about the reality of ice ages only began in earnest in the 1830s. The very first use of the term, as far as is known, was in a short, humorous poem written by a German botanist named Karl Schimper, who read and distributed copies of his little literary contribution to friends and colleagues at a scientific gathering in Switzerland in February 1837. Schimper was a brilliant but delusional scientist who was eventually committed to an asylum, where he died in 1867. He never became a formal participant in the debate about ice ages, nor did he produce any published works on the subject, but he was a close friend and colleague of the forceful and charismatic Swiss naturalist Louis Agassiz, who today is the person most closely associated with the formulation of ideas about a global ice age. Significantly, Agassiz was brought up literally in the shadows of the Alps, and glaciers—small mountain glaciers to be sure, but glaciers nevertheless—were part of the natural landscape of his childhood. By all accounts, Agassiz, a biologist whose first love was fossil fish, was a vigorous, highly intelligent, and very observant scientist. Like most of his contemporaries, he was initially skeptical about the claim that Alpine glaciers had been much more extensive in the past. But his conversion was rapid when he realized that many of the same landscape features that he observed being produced by contemporaneous mountain glaciers were also present far afield, in the ice-free valleys of his native country and even far beyond. Rural folk who encountered such features in their daily lives had reached a similar conclusion much earlier than Agassiz. The only way they could explain the large and exotic boulders they sometimes found plopped down in their fields was that they had been carried there by ice. That meant that in the past the glaciers must have extended far beyond their current boundaries.

As I hope will become apparent in this book, there is much that can be learned about the Earth, especially its climate, through careful study of the ice ages of the past. The story of how ideas about ice ages have developed, from the work of Agassiz in the 1830s to that of modern laboratories in the twenty-first century, is also a wonderful illustration of how science progresses: not on a smooth trajectory, but in fits and starts and sometimes even with "backward" steps, with long periods of accumulation of evidence and gestation of ideas, a certain amount of serendipity, occasional brilliant flashes of insight, and, especially in more recent times, technological advances. Perhaps because of the scale of the phenomena associated with ice ages, the subject has attracted its share of brilliant, charismatic, and eccentric characters, beginning with Louis Agassiz himself. A few are discussed in some detail later in this book: a self-educated Scot who made the connection between the Earth's orbit around the sun and ice ages; a Serbian mathematician who worked out—by hand, long before the advent of computers—a mathematical framework for determining temperature changes through time at any latitude on Earth; and an iconoclastic American schoolteacher-turned-academic who proved that parts of the northwestern United States had been ravaged by floods beyond imagining as ice age glaciers melted back into Canada.

Louis Agassiz began discussing his ideas about an ice age at scientific gatherings in 1837, and within a few years, in 1840, he had published his observations and theory in a book. What was truly radical about his treatment was his proposal that ice had covered most of Europe during the ice age, even, perhaps, most of the land on Earth. As is often the case with new concepts, this one did not initially win many adherents. However, the debate about the reality of ice ages quickly became one of the most fiercely argued controversies of nineteenth-century science. It continued, unabated, for decades.

And the eventual acceptance of the ice age theory was far from the end of the story. Since that time, literally hundreds, perhaps even thousands, of scientists have pursued research into the causes and effects of the ice ages, and many thousands of scientific papers have been written on the subject. In the course of that work, Agassiz's contributions have been remembered in small ways and large. When researchers discovered evidence of a vast ice-dammed lake that had formed along the margins of the melting ice age glaciers in the central part of North America, they named it Lake Agassiz. In Winnipeg, Canada, which lies within the area that had been covered by the waters of glacial Lake Agassiz, there is even an Agassiz microbrewery. Agassiz, who complained when he came to the United States about the American practice of drinking iced tea with lunch instead of wine, undoubtedly would have been pleased.

In principle, the idea of an ice age is a simple one—in the past, it was colder, glaciers were much more extensive than they are today, and huge ice sheets covered large sections of the continents that are now free of ice. However, understanding the phenomenon and determining how an ice age occurs, and what the ramifications are for the Earth and all its inhabitants, is far from simple. Today, it is difficult for anyone to be an expert in every aspect of ice age studies: the intellectual challenge presented by the geological evidence, with its multiple puzzles, has attracted the efforts of geologists, chemists, physicists, mathematicians, biologists, and climatologists. The work has taken on additional urgency in recent years because of mounting concern about the future of the Earth's climate system. While at first thought this might seem odd—the dominant problem today is global warming, not cooling—it has become clear that our planet has experienced huge climate shifts during the current ice age (as we shall see, the Earth today is still in the grip of an ice age). Understanding how these changes in the global climate occurred in the past, and what their effects were, is a key step toward predicting future changes. But in spite of the great advances that have been made in working out the details of what actually happened during the ice age, there is still much uncertainty about how, and especially why, an ice age actually begins. To be sure, there are hypotheses, but none have yet attained the status of an accepted scientific theory. Much remains to be done.

Louis Agassiz built his ice age theory within the framework of the then-popular catastrophist view of Earth history: the idea that rapid, large-scale events were responsible for many geological observations. He didn't really concern himself with a mechanism; he just assumed that temperatures had plummeted suddenly and the Earth "froze." He envisioned glaciers extending as far south as the Mediterranean Sea in Europe, and deep into North America. However, later research has shown that Agassiz's ice age was neither as rapid in onset as he proposed nor just a single cold period. We now know that the Earth's most recent ice age comprises a long succession of ice incursions deep into Europe (although not as far as the Mediterranean) and North America, separated by much warmer periods.

It is often not appreciated that today's climate is just a geologically short warm spell in this continuing ice age. But in addition to the ice sheets of Greenland and Antarctica, mountainous regions today sustain permanent ice fields even in the tropics. The brilliant white cap on Mt. Kilimanjaro described by Hemingway in *The Snows of Kilimanjaro* is actually a permanent glacier, in spite of the fact that Kilimanjaro is only 300 km (roughly two hundred miles) from the equator. The Andes too host equatorial glaciers. If you were an astronaut circling the Earth at the end of a northern winter, you would observe that nearly half the land area and more than a quarter of the oceans were white with snow and ice. Only a fraction of this is permanent glaciers, but still, about 75 percent of all the fresh water on our "blue" planet is frozen in glaciers. Even so, in comparison with the average of the past few million years, the present-day interglacial climate is benign. The last time the Earth was as warm as it is today was about 120,000 years ago; for most of the time since then it has been much, much colder.

All of the evidence we have about past climates suggests that the Earth has been progressively cooling for the past 50 or 60 million years.

Before then, most of the world had experienced warm temperatures—the fossil remains of tropical and subtropical plants and animals from those times are found even north of the Arctic Circle. Sometime near 35 million years ago, there was an especially sharp drop in global temperatures—this is when, most researchers believe, glaciers began to form in Antarctica. However, although temperatures continued to fall as the Antarctic icecap grew, it was not until about 3 million years ago that permanent glaciers appeared in abundance in the Northern Hemisphere, again accompanied by an abrupt temperature decrease. This is generally agreed to be the start of the current ice age, and since that time, most climate changes around the globe have been associated with the waxing and waning of ice sheets in the Northern Hemisphere. Fortunately for us, the glaciers have withdrawn to high altitudes and latitudes during the present warm period. But on average, for the past few million years, the Earth has been considerably colder than over most of its four and a half billion years of existence. During much of Earth history, except for short, rare, intervals, glaciers such as the one on Kilimanjaro have been absent. In contrast, within the current ice age, warm periods with moderate climates similar to the present have been short by geological standards, generally lasting only ten to twenty thousand years. We are already about ten thousand years into the current warm episode. If history is any guide, and if human activities don't warm the Earth too severely, the ice will return, and quite soon on a geological timescale. The sites of cities such as Montreal and Edinburgh and Stockholm, and perhaps even New York and Chicago, will be buried deep in glacial ice, as they were in the past.

You might reasonably ask: How do we know these things? How do we know that the Earth has been cool for the past few million years, compared to the rest of its history? How do we know that the Earth is still locked in an ice age characterized by a series of advances and retreats of ice over North America and Europe? One of the aims of this book is to answer these questions, and also to delve into some of the history behind the answers. In doing so, I hope also to illustrate the startling ingenuity of some of the scientists who have investigated such questions, and the deep curiosity about how the Earth works that has pushed them toward their goals. And also to show why such work is important for understanding—and perhaps even shaping—man's impact on our small planet.

Before embarking on this discussion, however, it is worth reviewing a few general aspects of ice ages. First is the meaning of the term itself, because its usage can be confusing. Initially, "ice age" referred to Agassiz's original concept, a period of cold at some fairly remote time in the past, during which most of Europe was covered with thick glaciers. As already noted, however, it was fairly soon discovered that there had been a whole series of "ice ages" that were really part of the same episode, separated by short warm periods during which forests grew and animals roamed over land that had once been buried deep in ice. Climate zones marched up and down the continents as the temperature changed and glaciers grew and then melted back again. We know this about the ice age of the past few million years because the fossil record of changing animal and plant species is quite well preserved. But there have been ice ages in the much more distant past, too—hundreds of millions and even billions of years ago. About those intervals we know much less, but they too probably had both cold and warm periods. In current usage, the term "ice age" properly refers to an entire cold episode, including its short warm periods. Hence, it's possible to talk about the "current ice age," which seems a contradiction in terms on a hot summer's day. Large-scale advances and retreats of ice during an ice age are usually referred to as glacials and interglacials respectively. Ice ages finally end when the cycles of these glacials and interglacials cease, and permanent ice, if it exists at all, becomes a minor feature of high mountains and polar regions. Our current ice age is often referred to as the Pleistocene Ice Age, taking its name from the subdivision of the geological timescale that more or less coincides with the time during which the Northern Hemisphere has experienced glaciation. Although the current ice age actually began somewhat before the start of the Pleistocene, I shall use that nomen-clature in this book as a convenient way to distinguish it from others in the geological record.

In addition to understanding the usage of the term "ice age," there is also the question of what, exactly, a glacier or icefield is, and how it comes about. Glaciers are actually nothing more than huge accumula-tions of snow. Pressure from the weight of overlying snow transforms those fluffy flakes that so delight children on a winter's day into the hard ice of a glacier, brittle enough to crack open in deep crevasses and strong enough to pluck solid rock from a mountainside and carry it down a valley. At the very surface of a glacier, the winter snow is as loosely packed as the drifts you shovel from your driveway after a storm (of course, that snow really doesn't seem to be so loosely packed when you're at it, but that's another story . . .). But dig down into a glacier, and you'll find the individual snowflakes—themselves just exquisitely formed crystals of ice—packed together much more tightly, the air between them forced out. Probe even more deeply, and you'll discover that the snow has recrystallized into a continuous mass of solid ice, as clear as the ice cubes in your refrigerator.

A bona fide glacier must be permanent. Generally, this implies that sufficient fresh snow must accumulate during the cold months to offset melting during the summer, although on a year-to-year basis, glaciers may expand or contract, depending on local and global climatic condi-tions. Today, most glaciers around the world are in retreat because of the warming climate, and it appears that the rate of melting is acceler-ating. This has been documented spectacularly in places such as the Alps, where historical records have been kept and dated sketches and photographs are available to compare with the present extent of ice. Even over periods as short as a few decades, satellite images show that dramatic shrinkage of mountain glaciers has occurred in the Andes, the Himalayas, and elsewhere. It is estimated that many small moun-tain glaciers will vanish completely within ten to twenty years unless there is an abrupt and unexpected change in the present warming trend.

The rapid melting of glaciers around the globe, while an ominous reminder of global warming, has been an unanticipated boon for archeologists. In 1991, climbers found the frozen body of a 5,300-year-old "Iceman" in a retreating Alpine glacier, complete with intact tattoos on his well-preserved skin. More recently, archeologists and biologists have begun making systematic surveys of melting glaciers in Alaska and northern Canada, not to monitor their retreat, but to retrieve the whole animals, human hunting implements, bones, and even the fresh-frozen animal dung that is disgorged as the glaciers melt back. It has become apparent that glaciers are invaluable storehouses of frozen materials from prehistoric times, containing clues about the animal species that were abundant in the past, what their diets were, and how native peoples hunted them. In 1999, melting ice in northern British Columbia yielded a human body that, although less ancient (only about 550 years old) than the Alpine Iceman, was also well preserved and accompanied by clothing and various tools and implements, all frozen in glacial ice. Named Kwaday Dan Sinchi ("Long Ago Man Found"), he has become the focus of intense interest on the part of both native peoples of the region and scientists. Analyzing DNA from humans and other species preserved in glaciers has the potential to open up whole new areas of biology and anthropology for investigation. Indeed, DNA samples have been collected from members of various First Nation tribes across Alaska, northern British Columbia, and the Yukon territory in an attempt to investigate possible links between present-day inhabitants and "Long Ago Man Found."

That glaciers are such rich sources of organic material has only recently been realized. It is one of those discoveries that is obvious only in hindsight—the animal and plant remains and waste that would decompose quickly in unglaciated regions are preserved, only to be released when the ice melts. This new knowledge should give pause to anyone wanting to slake his thirst with water that is "pure as the driven snow," fresh from the melting snows of a glacier.

Dead birds and animals aren't the only things stored in permanent icefields. The transformation of snow crystals into the hard ice of glaciers has also left us a remarkable and detailed record of the changing climate of the current ice age, reaching back through several glacial-interglacial cycles. Especially in Greenland and the Antarctic, where snow has been accumulating continuously for hundreds of thousands of years and the glacial ice is very thick, the history of local and global climate has been preserved, literally frozen into the ice, with high fidelity. Deep cores have now been drilled there, the precious samples wrested from the inhospitable poles of the Earth, whisked into refrigerators, and distributed to laboratories around the globe. The cores are layered with bands like tree rings, each band recording an annual cycle of summer warmth and winter snows. It is possible to count the layers back into history, a thousand years, ten thousand years, a hundred thousand years. In some of the layers, tiny particles of volcanic ash that have been carried by the winds to their final polar resting place record huge volcanic eruptions halfway around the world. Throughout the cores, the ice crystals themselves have a story to tell—they provide a continuous meteorological record because their chemical properties depend on the local temperature when they formed. And as the snow of thousands of years ago was gradually buried and recrystallized into ice, tiny bubbles of the ambient air were trapped. With care, these miniature time capsules can be retrieved from the ice cores and analyzed. It's a bit like opening Tutankhamen's tomb or finding an ancient petroglyph. The bubbles provide a glimpse into the past, making it possible to take direct measurements of the atmosphere as it was thousands of years ago—including, among other things, its content of the greenhouse gases thought to be responsible for global warming. From such analyses has come a detailed, and in many ways very sobering, record of how the Earth's climate has changed through the past few glacial cycles. As we shall see in this

Figure 1. The glaciers of the Pleistocene Ice Age have left their marks indelibly across much of the Northern Hemisphere. This polished, scratched, and grooved bedrock in southern Ontario, Canada, is a testament to the abrasive power of flowing ice and the rocks, gravel, and "grit" embedded within it. Photograph courtesy Professor Kenneth Hamblin, Brigham Young University.

book, some of the climate fluctuations documented from ice cores may have directly influenced the course of history, and possibly even the evolution of our species.

The tremendous power of the ice age glaciers is something that is not easy to imagine. Some inkling of their ability to shape the landscape can be gleaned from the icecaps of Greenland and the Antarctic, but most of us have never ventured to those remote regions and so have no way to gauge their effects directly. However, at more temperate latitudes, the deep grooves and scratches that the ice age glaciers left in solid bedrock as they crept across the land are often still visible, silent but very graphic testament to their abrasive power. At the base of a glacier several kilometers thick, similar to those that covered the northern parts of Europe and North America during the last glacial interval, the pressure is tremendous—roughly equivalent to that at a depth of several kilometers

Figure 2.   Glaciers shape the landscape on both small and large scales. This air photo shows a large expanse of land in the Northwest Territories, Canada, affected by Pleistocene glaciation. Long glacier-produced ridges are lined up parallel to the direction of ice flow, which was toward the bottom of the picture. By permission, Canadian National Air Photo Library. Copyright Her Majesty the Queen in Right of Canada.

in the ocean. Should a glacial interval recur in the future, the ice would not just bury northern cities, it would simply scrape them off the surface of the Earth and, eventually and quite unceremoniously, dump the twisted and mangled remains far to the south. Figures 1 and 2 give some idea, however imperfect, of just how powerful ice age glaciers are.

CHAPTER TWO

Fire, Water, and God

Louis Agassiz's theory of a global ice age overturned the conventional wisdom about the Earth's past. It also bruised a few egos, especially those of scientists who had built their reputations on quite different versions of the planet's history (in this respect, at least, scientific disputes seem not to have changed much over the intervening century and two-thirds). They did not appreciate this brash young newcomer—a zoologist at that—with his revolutionary ideas. Agassiz himself had anticipated opposition to his theory, but he was taken aback at the virulence of some of the critics. He saw the geological evidence for an ice age as overwhelming, and he could not fathom how others could interpret it differently. But many scientific advances have been similarly unpopular initially. Often it requires the genius of a conceptual thinker, which Agassiz clearly was, to launch a new way of thinking about natural phenomena. Just how clear—and, paradoxically, at the same time unnoticed—the evidence for glaciation was, even to naturalists who were accustomed to careful observation, is nicely illustrated by a passage from Charles Darwin's autobiography. In 1831, just a few years before the publication of Agassiz's ice age theory, Darwin was on a field trip in Wales with a prominent British geologist. Anxious to find fossils, they scoured hill and valley, examining the rocks in great detail. But they completely missed the evidence for glaciation that surrounded them. Years later and by then fully aware that the Earth had experienced extensive glaciation, Darwin wrote about his earlier field excursion: "On this tour I had a striking instance how easy it is to overlook phenomena, however conspicuous, *before they have been observed by anyone* . . . neither of us saw a trace of the wonderful glacial phenomena all around us; we did not notice the plainly scored rocks, the perched boulders, the lateral and terminal moraines."

The italics above are mine. They emphasize the common experience that is implicit in the phrase "point out the obvious." Some things are invisible until someone shows them to you; then they pop up everywhere. Darwin marveled that he and his geological colleague could have overlooked these glacial features that he later found so conspicuous that "a house burned down by fire did not tell its story more plainly than did this valley." But it is also true that it would have required Darwin and his colleague to make a huge leap of understanding to interpret the glacial features correctly. Agassiz argued for a global-scale ice age by extrapolation from observations he made studying glaciers in the Alps. The true extent of the two great continental ice sheets that still existed on the Earth, in Greenland and the Antarctic, was still unknown. It was to be more than a decade before explorers established that a single massive icecap covered Greenland. With no well-known natural analog, it was difficult for most people to imagine the magnitude of the phenomenon that Agassiz proposed, and even harder to envision glaciers in places like Wales, with no high mountains and more than a thousand kilometers from even the small Alpine glaciers of Switzerland.

The ice age theory came to prominence in the 1830s. Historians of science often refer to the final decades of the eighteenth and the early decades of the nineteenth centuries as the "heroic period" of geology. Thousands of people—obviously very few of them actual practitioners of the subject—regularly turned out to hear lectures by the geological superstars of the early 1800s. The Geological Society of London, founded in 1807, had concerns that too many people would apply to join. Considering that the middle class, although expanding rapidly, was still small, and the traditional educated elite smaller still, these facts speak impressively of a pervasive interest in the Earth and its history. Geology was, in its early days, an activity in which almost anyone with time and the means to travel a bit could participate, and even contribute to. Not only that, it offered an opportunity to get out into the countryside, away from the rapidly industrializing and heavily polluted cities.

This was also a time when European (and North American) philosophy, science, and society were undergoing rapid change. The changes had actually begun a century or more earlier, but the pace was accelerating. It was a time when science, based on mathematics, experimentation, and observation using new instruments such as microscopes and telescopes, began to challenge the status quo. During the seventeenth and eighteenth centuries, momentous advances were made in understanding the "universe," a term that at the time encompassed virtually all of natural science. But it is worth remembering that the new discoveries were being made, not through a scientific enterprise anything like that of today, but largely by individual geniuses, some of them in universities, some in government, some independent inventors. They were "natural philosophers," not scientists in the present-day sense of the word. Their philosophy, their view of the world, was a radical departure from earlier philosophies, which tended to be based in religion and founded on decree rather than reason. Galileo was interrogated by the Inquisition for his heretical proposal, based on observation, that the Earth moved in an orbit around the sun and was not at the center of the universe. But even giants such as Newton, while seeking truth through observations and mathematics, were at pains to point out that their work was consistent with, not antithetical to, a theological or scriptural foundation for the world, and especially for man.

The natural philosophers of the seventeenth and eighteenth centuries corresponded regularly with one another and met to present their work at gatherings such as those of the Royal Society in London.

Through networks of people with similar interests, their ideas gradually spread and gained prominence. There was an understanding, at least in intellectual circles and among the educated segments of society, that one should be skeptical of dogma, and that logic and reason could solve many problems in all realms of life. However, by the beginning of the nineteenth century, the century that would see the idea of ice ages proposed and eventually accepted, there was a considerable backlash against many of the tenets of the Age of Reason. The idea that nature, man, and indeed the universe were governed by a set of physical laws was a concept that was simply too stark and inhuman for many people to accept. The thought that there might not be any mystery, that there might not be a Divine Providence backstage pulling the strings, was contrary to everything that most Europeans believed. They were also concerned that commercial and practical concerns, rather than spiritual ones, were becoming the driving forces for their governments, and that church was being separated from state—formally, in the case of the newly independent United States. On top of that, Europe and North America were in the throes of an Industrial Revolution that was sending shock waves through society and changing the way thousands, if not millions, of people worked. There was also political turbulence: first the American and then the French Revolution had overturned the old order. One of the outcomes of all this turmoil was an upswing in religious conversions and the founding of a rash of new religious sects.

How did the young science of geology, which, more than most sciences, intersected with long-held religious beliefs, fit into this turbulent mix of cultural change? And how were the developing ideas about ice ages affected? This is not the place for a detailed history of the field, but some background is necessary to put the work of Agassiz and his forerunners in context. As the title for this chapter hints, there were a few prominent themes that guided ideas about the Earth in the early days of natural science and geology. One was spiritual, the idea that there was a creator who had constructed the Earth and everything on it. In Europe and the Americas, this evolved from a strict adherence to the six days of Creation described in the Bible to much looser interpretations as more and more geologic evidence accumulated that was contradictory to the biblical story. Then there was a long-standing controversy about the origin of the rocks in the Earth's crust: had they all been laid down in some primordial ocean, the land gradually emerging as the seas withdrew (this scenario too had religious overtones, because it was sometimes linked to the biblical Flood) or was there internal heat and fire that fused minerals and made the materials that we today call igneous rocks?

In Italy and in France in the eighteenth century, and even earlier, there were men who, usually through their natural curiosity rather than their vocation, made perceptive and prescient observations about the nature of fossils, the existence of extinct volcanoes, and how sedimentary rocks must have formed. They published their work, but for whatever reason, it was not widely known in Britain, which was to become the acknowledged leader in the development of geology as a modern science in the nineteenth century. There, especially, religion continued to have a very strong influence on science during this entire period. But throughout Europe in these early times there was often not much distinction made between the study of nature and theology. The underlying assumption was that there was a God who had created the Earth and all living things. That meant that human history and the Earth's history were coincident. Observations of the natural world had to fit into this framework.

To a present-day reader, many of the early books about the Earth are fantastical conjectures without much basis in reality. Some attempted simultaneously to explain facts about the present Earth and still conform to the biblical story of genesis, the details of which were widely known to people of the time. But the more influential of these accounts were essentially philosophical enquiries, steeped in knowledge of the writings of "the ancients," especially the Bible, but also seeking "truth" in the spirit of the Age of Reason. Perhaps the most important of these was written by Thomas Burnet, a British academic who was first and foremost a theologian—he served for a time as a chaplain to King William III. His book, written in Latin, appeared in 1681; the English version published a few years later had the title *The Sacred Theory of the Earth,* and it contained chapters dealing with topics such as Paradise, the Flood, and the Conflagration. Burnet likened the primordial Earth to an egg—slightly oval in shape, with a smooth and "pure" surface, and, in its structure, layered in several "orbs," which he equated with the yolk, white, and other parts of an egg. However weird all of this seems to us now, Burnet's book was a best-seller in England and abroad, and it affected thinking about the Earth for the next century. Writers and poets—including Coleridge and Wordsworth—acknowledged its influence. Newton, a contemporary, thought that Burnet gave "the most plausible account" of the presence of seas, mountains, and rocks on the Earth.

Nevertheless, Burnet's book was controversial. He tried, he explained, to put some science into his theory. At the same time, if God had created the Earth, he could rationalize that it was also a theological enquiry, because learning about the Earth was a route to knowledge of God. Burnet was a philosopher and a theorist, rather than a close observer of nature, but he tried in his writing to bring some sort of order to the often conflicting scenarios that characterized various contemporary ideas about the Earth's history. He constructed a theory that could be held up to nature for verification. However, both scientists and theologians took issue with much of what he wrote. They paid particular attention to his version of the biblical Flood. Burnet calculated that there wasn't enough water in the ocean to cover the land to the prescribed depth, and he concluded that the Flood could not have been caused simply by heavy rains. And so he devised a different, somewhat more complicated explanation. The problem for many theologians, however, was that his was a physical explanation, rather than one based on God's displeasure with man. Burnet envisioned the early Earth—the paradise phase—not only to be smooth like an egg, but to be blessed with perpetual summer. For an Englishman, this would surely be a vision of paradise. Under these conditions, he wrote, within a few hundred years the Earth's crust would start to dry out and crack. Not only that, but the region under the crust—the albumin in Burnet's egg analogy—was mostly water and would be heated up by the constant sunshine. The deepening cracks, combined with the expansion of the heated water, would eventually cause the crust to collapse in a violent paroxysm, the interior water shooting high into the air and covering the entire surface. According to Burnet, the water would have slopped around the Earth's surface for months, like water in a shaken bucket.

This deluge would have left "ruin" in its wake—that is, the landscape of the seventeenth century familiar to Burnet and his readers; valleys, mountains, and seas replaced the smooth surface of paradise. Rugged topography was imperfection; mountains especially were dark and inhospitable and possibly evil. "Thus perisht the old World, and the present arose from the ruines and remains of it" wrote Burnet. And later in his narrative: "And so the Divine Providence, having prepar'd Nature for so great a change, at one stroke dissolv'd the frame of the old World, and made us a new one out of its ruines."

Religion continued to influence investigations of the Earth and its history for the century and a half that would elapse between the publication of Burnet's book and Agassiz's theory of ice ages. In Britain, even in the early 1800s, most of the people we now consider early geologists—and all of those who were educating students in geology at the universities—were actually clerics, in the employ of the church, although they probably spent more time on their geological pursuits than their theological ones. But there were other, secular, influences as well. The practical concerns of miners, engineers, and navigators required ever more accurate information about minerals, the nature of coal and metal deposits, and the Earth's magnetic field. Government funding began to flow into some areas of science as entrepreneurial scientists realized that they could finance their investigations by emphasizing the economic payoffs. The making of precise instruments for measuring physical properties of the Earth became highly developed.

The earliest geologists naturally enough concentrated most of their attention on the land surface, because it is the most easily accessible part of the Earth. It is also where the evidence that eventually led to the development of the theory of ice ages lay, in the shape and nature of the glaciated landscape. Because the peak of the most recent advance of ice in the Northern Hemisphere occurred only 20,000 years ago—a geological blink of the eye—the evidence is abundant. There has not been time for it to be eroded away or destroyed by the mountain-building forces of plate tectonics. The evidence for more ancient ice ages is much less apparent. Had no ice age occurred on the Earth for hundreds of millions of years, there would have been no glaciers in Switzerland in the nineteenth century, and none of the geologically ephemeral evidence that was used to document the ice age theory would be present. We might even today be puzzling over the significance of the much more cryptic evidence that exists for the Earth's earlier ice ages.

Much of the early geological exploration was focused on questions of how the various rock types observed at the Earth's surface had formed, or how the morphology of the landscape had originated and evolved. As the sway that religion held over explanations for the natural world waned, most of the men taking part in these investigations—they were all men—began to acknowledge that if God were involved at all, it was through the action of secondary agents or processes. Initially, long before there was any hint that ice was important in sculpting the land, the debate centered on whether the rocks at the Earth's surface were deposited from water or somehow formed in the heat of great fires in the Earth's interior. At the poles of this debate were three men. Abraham Werner (1749–1817), a Prussian scientist who worked at the Freiberg Mining Academy in Germany for most of his professional life, formulated a comprehensive scheme in which he proposed that virtually all rocks on the Earth had been deposited by the waters of the ocean. In his view, they were either direct chemical precipitates from the sea or mechanical deposits of debris washed in from the land. At the same time, in France, Nicolas Desmarest (1725–1815) realized that, at least in some places, quite different processes had been at work. Desmarest was a remarkable civil servant who traveled widely in France in the course of his government business. He was also obsessed with geology. He knew that there were active volcanoes in some places, and that their molten products cooled to form surface rocks. But as he traveled in the southern and central parts of France, he found volcanic rocks that were very far from any centers of active volcanism and made detailed surveys of them. He realized that they must have been formed at some time in the past, and that the volcanoes responsible for these rocks had since become dormant. Finally, James Hutton (1726–97), a Scottish intellectual who, like Desmarest, was a keen observer, realized that volcanic lavas erupting on the surface aren't the only rocks produced by great heat. Based on his field observations of granite and related rocks, he concluded that they had once been molten, but that instead of flowing out onto the Earth's surface from a volcano, these materials had cooled and congealed while still deep within the Earth. Hutton realized that an internal source of heat was required. He was also one of the first to recognize the cyclic nature of geologic processes— hills and mountains are worn down by erosion, the debris produced by this process is deposited as sediments in the sea, buried sediments are heated and fused together to make rocks, and, finally, to complete the process, they are thrust up again by some force to make the hills and mountains of the continents. It was Hutton, contemplating the vast amount of time required for these geological processes, who crystallized for many the immensity of geologic time with his famous line that there is "No vestige of a beginning, no prospect of an end."

Although Fire—the Demarest and Hutton views—eventually won out, Werner and his many students, dubbed the Neptunists because of the importance of water in their theories, had a great influence on the young field. A highly respected British cleric and geologist, Adam Sedgwick, writing about the controversy in the nineteenth century and attempting to inject a little humor, said "For a long while I was troubled with water on the brain, but light and heat have completely dissipated it." And while Werner himself never tried to equate his primordial ocean with the biblical Flood, it was a seductive idea for those who still tried to connect all aspects of the natural world to a strict interpretation of the Bible.

The ideas of the Neptunists, even though partly discounted by the early 1800s, strongly influenced many of the opponents of Agassiz's ice age theory. Glacial deposits are often just heaps of boulders and gravel, or sometimes banks of sand and silt, carried away from the front of a retreating glacier by meltwater, and they were invariably interpreted by early observers as water-deposited sediments. Critics of the ice age theory continued to promote this view. They had problems explaining how a flowing stream could leave behind both large boulders and small sand grains simultaneously, but such difficulties were glossed over. They simply discounted Agassiz's contention that the deposits were the work of glaciers. They bolstered their argument that the glacial sediments must have been deposited in one or more huge floods by pointing to their often chaotic character. When it was shown that many of the boulders were simply too massive to have been carried by water, a new twist was added to the old ideas. Perhaps the boulders *had* been carried by ice, it was said, but icebergs floating south from the polar region, not solid ice advancing directly over the continents. Sailors, especially fishermen and whalers, were well acquainted with icebergs in the North Atlantic. If sea level had been higher in the past, or if the land had been depressed, one could imagine melting icebergs dropping their rocky burdens onto the submerged English or European countryside.

Such was the social and scientific backdrop when Agassiz's theory that there had been a worldwide ice age emerged on the scene. Remnants of Neptunist ideas and thoughts about the biblical Flood still influenced some naturalists. That the debate about large-scale glaciation went on for so long, at least three decades, is a lesson in the durability of ideas—even when there is very strong evidence that they are wrong.

CHAPTER THREE

Glaciers and Fossil Fish

Louis Agassiz grew up in Switzerland, in a village that was almost surrounded by water—two lakes and a river flanked his little town. From
the time he was a small boy, he loved to go fishing. Local fishermen
would take the parson's son out in their boats and teach him their
secrets. They liked this precocious lad, and they soon realized that he
was a quick study. Although none of them knew it at the time, fish
were to be an important part of Agassiz's later life. So too were the glaciers of the Alps, looming on the horizon not far from his home.

Agassiz was to become one of the nineteenth century's most
respected naturalists, a man whose rapid rise to prominence was the
result of a unique combination of intelligence, singular determination,
force of personality, and perhaps a certain amount of luck. But his personality was certainly central to his accomplishments. The village fishermen who took him out in their boats weren't the only ones to succumb to Agassiz's spell. He had an almost instant effect on everyone he
met, and throughout his life he was able to persuade people to help
him—with money, with sketches of his specimens, or with permission
to use their libraries or inspect their fossil collections. Quite often their
offers were spontaneous, and even complete strangers did not escape his
appeal. When Agassiz was in his late teens studying at Zurich, he and

his brother once traveled home on foot, a distance of about a hundred and fifty kilometers—almost a hundred miles—as the crow flies, and much farther for the walkers. Along the way, they were offered a lift by a well-to-do citizen, who, during his brief encounter with the boy, was so impressed that he later wrote to Agassiz's parents saying that he would be happy to pay for the entire cost of Louis's education. Although the family didn't accept this generous offer, the story illustrates just how powerful Agassiz's enthusiasm and charm could be, even when he was a young man. The incident also foretold a common occurrence in his later life—financial rescue by a wealthy patron. Agassiz paid scant attention to money, and his dreams and ambitions in science often far outstripped the conventionally available resources.

The Agassiz family envisioned a traditional middle-class life for their son—a profession such as medicine or business that would command respect in the community, marriage into a good family, and a comfortable life at home in Switzerland. But it was not to be. In spite of their best efforts, Agassiz was unwavering in his determination to become a naturalist. He was not a rebel in the conventional sense, and he always had great respect and love for his parents, but he also always managed to persuade them—either himself or through influential relatives or mentors—to do things his way.

One of the first instances of this characteristic that we know about occurred when Agassiz, at age fifteen, had finished the first stage of his education at a nearby school. He had impressed his teachers with his learning, especially his gift for languages. In his spare time, he fed his insatiable curiosity by collecting and learning about everything he could lay his hands on from the natural world—insects, plants, animals, fish. He seemed to be on an academic trajectory. But his parents had a different plan: now that he had had some education, they would send him to nearby Neuchâtel, where he could serve an apprenticeship with one of his uncles, who ran a business there. Young Louis would thus learn the intricacies of commerce. But in fact he had not the slightest interest in doing so. Although his vision of a career was still hazy, he knew he wanted to continue his education. He wanted to be a "man of letters," he wrote in a private note to himself; he wanted to "advance in the sciences." Shrewdly for a lad of fifteen, he enlisted the help of one of his teachers, who spoke to his parents about their son's future. Before long, they concurred with Agassiz's wishes, although they may not have realized who was really behind the plan. Louis was sent off for two years of additional schooling in Lausanne—a temporary delay, his parents thought, in his entry into the world of business. The two years in Lausanne, however, only strengthened Agassiz's resolve to become a man of science. Although he was there ostensibly to study the humanities, he attended lectures on natural history, spent time in the natural history museum, and learned anatomy from a relative who was a physician in the town.

By then there was no turning back. Agassiz never did return to Neuchâtel to serve an apprenticeship with his uncle; instead, he went on to study in Zurich, and from there to Heidelberg and Munich. It was in Munich that one of his professors asked him to work on a collection of fish from the Amazon that had been collected during an expedition some years before, but never described or cataloged. Louis readily agreed, and with characteristic energy, and in spite of the fact that he was attending lectures and separately doing his own studies of European fish, he completed the work far ahead of schedule. He published a book on his investigation, *Brazilian Fishes,* in 1829, which was received with considerable acclaim. Agassiz was only twenty-two and still a student. Already he had entered the world of scholarship and had come to the attention of naturalists throughout Europe.

How, then, did this brilliant young naturalist whose specialty was fish end up being forever identified with the concept of ice ages? More than a little serendipity was required. But it should also be remembered that this was a time long before the era of narrow specialization in the sciences, when "naturalists," especially, tended to be generalists who could and did pursue any phenomenon that piqued their curiosity. It was a time of exploration, of great general interest in the natural world, of expeditions to unexplored places to collect specimens for Europe's burgeoning museums. Agassiz's initial interest in glaciers was almost certainly stimulated partly by the challenge of deciphering the enigmatic landscape features of his native Switzerland but also partly by simple curiosity. ("Among all nature's phenomena, not a single one seems to me to be more worthy of the interest and curiosity of the naturalist than glaciers," he wrote in 1840.) His theory of ice ages was, moreover, soon woven into the framework of his ideas about the origin of life on Earth. Eventually, it became one of the pillars of his opposition to the idea of Darwinian-style evolution. It is ironic that this theory, formulated early in his career and truly a triumph of observation and deduction, was later to become an important part of his dogmatic and very nonempirical rejection of the developing ideas about evolution.

Serendipity, chance observations or opportunities, and unexpected results are common enough in science even today, and they played an important role in Agassiz's career. His route from being a student of zoology and medicine in Munich to authoring the theory of ice ages was in some ways unremarkable and in others quite amazing. (Notice that Agassiz studied both zoology and medicine. His parents were by this time resigned to the fact that he was destined to be a naturalist, but they urged him to complete a degree in medicine so that he could always work as a physician if a career as a naturalist didn't work out. Ever dutiful, Agassiz followed their advice—but it was zoology, not medicine, that really captured his heart and mind.) When he had completed his book on Brazilian fish, Agassiz decided to dedicate it to the great French naturalist Baron Georges Cuvier. Agassiz had never met Cuvier, but he idolized him and confessed when he sent him a copy of *Brazilian Fishes* that "your works have been till now my only guide." Still, the dedication may have been made with one eye on the future. Agassiz's letter also laid out his hopes and plans for a career in science. Cuvier replied, which flattered and encouraged the young naturalist. And a few years later, in late 1831, shortly after he had completed both his degree in medicine and a Ph.D. in zoology at Munich, Agassiz was in Paris seeking an audience with the great man. They met, and Cuvier, like so many others, was quickly won over by Agassiz's intellect, enthusiasm, and complete dedication to his work. So impressed was Cuvier that before long he had turned over to Agassiz one of his own projects, one that seemed tailor-made for the young scientist: a comprehensive examination of the entire fossil fish collection then housed at the French National Museum of Natural History. It was the kind of large, important, and, if done properly, reputation-enhancing study that could fully engage Agassiz's interest. He waded into it with gusto, and for a short but intense time, he worked together with Cuvier, learning about the importance of careful observation and the intricacies of reconstructing anatomy from fragmentary fossil evidence. He also absorbed Cuvier's theories about the origin of life, which had been developed from long and careful study of fossils: that animals could be divided into several groups with no connections among them; that species were "fixed" and did not change; and that there had been periodic catastrophes in the Earth's history that had wiped out most living things, with newly emerging species bearing no relationship to those that preceded them. This, of course, was a very different scenario from the one that would soon be proposed by Charles Darwin. However, it was one that undoubtedly played a part in the development of Agassiz's ideas about a catastrophic ice age. Cuvier was a devout man, and he believed that the conclusions he reached from the study of fossils were simply manifestations of a higher plan, God's plan. For the rest of his career, Agassiz was to work within a similar framework, believing that he was revealing the creator's design through his studies of nature.

Cuvier's accomplishments and position made him an influential man in France, with important connections in science and government. One of these was Alexander von Humboldt, another great man of the times, who happened to be in Paris on official business for the king of Prussia when Agassiz arrived to begin work with Cuvier. Humboldt was a renowned explorer, and, like Cuvier, a naturalist of high distinction. He was also interested in discovering and encouraging new talent.

Humboldt knew of Agassiz's book on Brazilian fish, and when he learned more about Agassiz's work from Cuvier and others, he invited the young scientist to his Paris headquarters. Soon he too was charmed by Agassiz's enthusiasm and impressed with his dedication to science. In turn, Agassiz had his eyes opened to the possibility that distinguished scientists could also move easily in the highest circles of society and government and have influence far beyond their chosen field—a lesson that most probably played a part in his assumption of just such a role many years later when he emigrated to the United States.

Agassiz was fortunate to have Cuvier as a guru, but their work together was short-lived: Cuvier died in May 1832. In that short time— Agassiz had been in Paris for only half a year—Agassiz had already come to regard Cuvier as both mentor and friend, and their brief association would influence him throughout his career. In spite of Cuvier's death, Agassiz was determined to complete the work on fossil fish that had been entrusted to him, but he was at a loss about how to proceed. His mentor was gone, and his personal funds were all but exhausted. But a guardian angel appeared just when Agassiz needed it the most. Unsolicited, Alexander von Humboldt sent Agassiz a check for a thousand francs so that he could continue the work. More than that, Humboldt was instrumental in securing a position for Agassiz that would, for the next thirteen and a half years, allow him to do his best scientific work—including his seminal studies of glaciers and development of the ice age theory. As fate would have it, the town of Neuchâtel, where Agassiz's parents had originally wanted him to serve an apprenticeship in business, was about to establish a new academy and a natural history museum. Equally important—at least as far as Agassiz's career was concerned—the canton's affairs at that time were governed jointly by Switzerland and the king of Prussia. Humboldt quickly mobilized support for Agassiz among the Prussian authorities, and he also wrote to the local Neuchâtel aristocracy in praise of this native son of Switzerland, the local boy who had already gained international prominence as a naturalist. They had little choice; how could such entreaties be ignored? And wouldn't Agassiz and his already substantial collections bring fame to their museum and town? Their response was to create a position for Agassiz at the new college—they appointed him professor of natural history and raised money for a modest salary. In addition, Humboldt arranged for the Prussian government to subsidize purchase of Agassiz's personal collections for the new museum, a move that helped him considerably financially. In the fall of 1832, at the ripe old age of twenty-five, Agassiz returned to Switzerland to take up his professorship. He was brimming with ambitious plans for the new college and museum, and also for his own place in the wider world of science.

It had always been apparent that Agassiz was a gifted communicator and an engaging companion at the café or pub, but it was in Neuchâtel that his skills as a teacher became obvious to everyone. He was an avid believer in what would be called, in today's jargon, inquiry-based learning. Within months of arriving at his new post, Agassiz had formed a local natural history society. In the beginning, it consisted mostly of the higher echelons of Neuchâtel society, but soon Agassiz had, it seemed, persuaded half the townspeople that they should be out collecting specimens and learning about zoology, geology, and botany. To be sure, he gave conventional lectures in his role as professor, but learning by doing was what he espoused most enthusiastically. Later, he would transform science education in America using the same approach. But for the moment, the whirlwind that was Agassiz had descended on quiet little Neuchâtel, and at least while he remained, it was never quite the same. He soon became a source of civic pride for many of the town's prominent citizens, and the success of the man and his museum became their personal concern. For scores more, he was an inspiring leader of field trips and teacher of nature. Not incidentally, the steady stream of fish, fossils, birds, and animals collected by these new nature enthusiasts allowed his small museum to grow rapidly.

But Agassiz had other things on his mind as well. He married a German sweetheart of many years and settled into a comfortable domestic life. He also threw himself again into the work on fossil fish that had begun with Cuvier in Paris. One of the main reasons he had taken the post in Neuchâtel, far from the great European centers of learning, was that it both offered him the opportunity to spend much of his time on his own research and provided him with a small but dependable income. It was a wise decision. Over the next decade, his research resulted in a series of volumes on fossil fish that were unequalled in their clarity of description and classification, and that were accompanied by beautiful, accurately drawn illustrations. His reputation grew steadily and rapidly, and his work brought widespread acclaim and a series of medals and prizes.

One summer during his stay in Neuchâtel—the summer of 1836—Agassiz took his family on holiday to a picturesque region of the Swiss alps near the town of Bex. The town was home to a well-known geologist, Jean de Charpentier, who was director of the salt-mining operations there. De Charpentier was a convivial man who frequently hosted scientists, naturalists, and other men of learning at his home, and he had urged Agassiz to visit. The two families got along well; de Charpentier's wife, like Agassiz's, was German, and Agassiz had rightly perceived that his own spouse would welcome the company of a compatriot after the relative insularity of Neuchâtel. The decision to spend that summer holiday in Bex proved to be a crucial one in Agassiz's career—once more serendipity at work. For Charpentier was convinced that the glaciers of the Alps had once been much more extensive, and he was keen to show Agassiz the evidence.

Charpentier had been brought to this conclusion through the work of a friend and colleague, an engineer named Ignatz Venetz. During his work in the Swiss countryside, Venetz had had ample opportunity to observe the so-called erratic boulders that bore no resemblance to local rocks and were often stranded far up on valley walls. Some of the boulders were scratched and faceted, features reminiscent of those he had observed on rocks at the very margins of active glaciers. He also saw arcuate ridges of boulders and gravel curving across the green Swiss valleys. They were almost identical to the mounds of rock debris—moraines—that marked the sides and ends of contemporary glaciers in the Alps. Venetz concluded that these features could best be explained by glacial action in the past, which meant that the Alpine glaciers must have been much more extensive. He had developed these ideas over a long period of time and had formally proposed them in 1829, some seven years before Agassiz visited Bex. Venetz also knew that similar features had been reported from other parts of Europe, and he concluded that glaciers had probably existed in those regions as well. He convinced his initially skeptical friend de Charpentier of the reality of the proposal by taking him out into the field and showing him the widespread and abundant glacial features. By the time Agassiz visited in 1836, de Charpentier had accumulated and cataloged evidence of extensive past glaciation both in Switzerland and elsewhere. But neither Venetz nor de Charpentier aggressively pushed their ideas about glaciation. Perhaps they simply did not realize the significance of their observations. For whatever reason, they were content to discuss their ideas in a low-key way with other scientists, and, for those willing to make the effort, to show them the field evidence. Many naturalists, especially those interested in natural history, knew about their views but did not give them much credence.

Agassiz was one of those familiar with the ideas of these two men, but until his summer visit to Bex, he too had dismissed their theory as unlikely. The holiday was a pleasant one—the scenery was idyllic, and the Agassizes found that they had much in common with the de Charpentiers. Most evenings would find Agassiz at de Charpentier's table with Venetz and others, enjoying their host's hospitality and engaging in long conversations about natural history and philosophy. We shall never know exactly what transpired during these discussions, but given de Charpentier's and Venetz's interests, they must have included debates about glaciers and their past behavior. Agassiz listened to what his colleagues had to say, and undoubtedly these evening brainstorming sessions influenced his thinking. However, with his strong belief in observation and hands-on science, it was almost certainly the visible field evidence that really convinced him that Venetz and de Charpentier had discovered something important. When they took him out into the Alpine valleys and showed him the moraines, erratic boulders, and glacial scratches and grooves, it was a revelation. When he returned to Neuchâtel that autumn, he was like a blind man suddenly given sight: he saw signs of glaciation—especially erratic boulders and glacial scratches—everywhere around him. And, together with his long-time friend from student days, the botanist Karl Schimper (the same Schimper who coined the phrase "ice age," and who was now working with Agassiz in Neuchâtel and living in the family home), he synthesized these observations and soon came up with a theory. With the zeal of a convert, Agassiz took up the cause of past glaciation and did his mentors one better: he proposed a period of *global* frigidity in the past, not just one in which glaciation in the Alps and a few other regions of Europe had been more extensive.

The announcement of Agassiz's ice age theory came only a year after his sojourn with de Charpentier in Bex. In 1837, the Natural History Society of Switzerland met in Neuchâtel, and Agassiz, as president of the society and host of the gathering, gave the introductory address. The delegates, who expected that Agassiz would discuss fossil fish or one of his other biological interests, were in for a surprise. After expressing his pleasure in welcoming them to Neuchâtel, and extolling the advances that were occurring in the sciences, Agassiz said that he wished to focus on a topic appropriate to the location: glaciers, moraines, and erratic boulders. Carefully acknowledging his debt to de Charpentier and Venetz, he laid out a comprehensive ice age hypothesis that presaged his book *Studies on Glaciers,* which would be published three years later. The address was the pivotal event that brought the idea of an ice age to the full attention of scientists. In the context of what had gone before, it was a grandiose scheme, and it generated instant and long-lasting controversy. What was truly new was Agassiz's proposition that during the ice age, a great sheet of ice had extended from the North Pole to the Mediterranean, before the Alps had even been formed. This was very different from the idea that Alpine glaciers had extended a bit further down their valleys in the past. It took even his friends Venetz and de Charpentier by surprise. Furthermore, Agassiz brought zoology into the picture by proposing that the ice age had extinguished many of the Earth's living creatures. And implicit in his theory was the idea that there had been significant climatic variations in the Earth's past. The conventional wisdom at the time was that the Earth had been cooling since its creation. Agassiz suggested instead that each geological period (already geologists had subdivided the Earth's past into different periods based on the fossil record) had had an equable, stable, climate, but was terminated by a frigid episode, after which temperatures recovered, albeit perhaps not quite to their previous level.

One imagines that many in the audience rolled their eyes. The scope of this scheme was too much even for de Charpentier, who was the person most responsible for convincing Agassiz that Alpine glaciation had once been much more extensive. Agassiz, even in his enthusiasm for this new (for him) subject, recognized that there was likely to be some adverse reaction. Toward the end of his address (which was later published) he said:

> I am afraid that this approach will not be accepted by a great number
> of our geologists, who have well-established opinions on this subject,
> and the fate of this question will be that of all those that contradict
> traditional ideas. Whatever the opposition to this approach, it is
> unquestionable that the numerous and new facts mentioned above
> concerning the transportation of boulders, which may easily be studied
> in the Rhône valley and in the vicinity of Neuchâtel, have completely
> changed the context in which the question has been discussed up to the
> present time.

And in this respect he was right. His theory was not a flight of fancy; it was carefully based on detailed field observations. Some of the observations he discussed drew on the work of de Charpentier and Venetz, but many were his own. The observational skills he had learned from Cuvier in Paris served him just as well in glaciological as in fossil studies. "Fortunately, in scientific problems, numerical majorities never settled any issue at first glance," Agassiz told the skeptical delegates, confident of his ability to win over critics in the long run.

Three years after his lecture at the Neuchâtel meeting, in 1840, Agassiz published his carefully compiled evidence for widespread past glaciation in a large volume written in French and titled *Études sur les glaciers*—translated as *Studies on Glaciers*. It was the formal presentation of the ice age theory to the world, and it is a remarkable book, truly a tour de force. Although he never considered studies of glaciers to be his primary scientific focus, in practice, Agassiz devoted a great deal of his time to this work, spending the better part of each summer in the Alps doing glaciological fieldwork. In his book, in engaging language, Agassiz described in great detail the observations that he and his colleagues had recorded during those summer field seasons: the temperatures, the nature of the crevasses, the morphology of moraines, the details of the grooves and scratches on the underlying rock, and much more. He included beautiful, if somewhat unnatural-looking, sketches of many of the glaciers they studied, often with tiny people—walkers, women in peasant dress, picnickers—or farm animals drawn in. After cataloging the field observations, and noting how similar features occur far from the present-day glaciers, Agassiz, in a few short sentences, laid out his revolutionary conclusion:

> In my opinion, the only way to account for all these facts and relate them to known geological phenomena is to assume that . . . the Earth was covered by a huge ice sheet that buried the Siberian mammoths and reached just as far south as did the phenomenon of erratic boulders. . . . It extended beyond the shorelines of the Mediterranean and of the Atlantic Ocean, and even completely covered North America and Asiatic Russia.

The frozen Siberian mammoths to which Agassiz refers in this passage had caused a great stir in Europe, and they featured significantly in the ideas he developed about the biological effects of the ice ages. Several reports that had filtered out from arctic Siberia described these giant mammals melting out of decomposing ice, almost perfectly preserved. Hair, skin, and flesh were intact—in fact, polar bears fed on the thawing animals and local villagers hacked off meat for their dogs. To Agassiz this indicated that the ice age had begun suddenly and had been a biological catastrophe. He also mentions here erratic boulders, the same kinds of boulders that had played a major role in convincing Venetz and de Charpentier about past glaciation. Erratics had puzzled geologists for decades and, as has already been mentioned, were at the focus of the debate between those who wanted to explain most geological features as originating in the biblical Flood and those who sought more natural explanations. Some of the erratics in the Alps are huge, weighing thousands of tons, and they are unusual because they bear no resemblance to other rock types in their immediate vicinity. Their great importance in Agassiz's theory was that they are markers of the extent of ice age glaciers.

Agassiz dedicated his book to Venetz and de Charpentier. Still, they were slighted—Agassiz did not consult them about his ideas, and they felt that his grandiose theory misrepresented some of their ideas. No one before had postulated a truly global cold period, and no one was as enthusiastic or confident about promoting the theory as Agassiz. In addition to showing how important ice is in shaping the landscape, Agassiz also introduced the idea that drastic climate change had occurred in the Earth's history, and that the cold of the ice age had strongly affected life on Earth. Each of these was a new idea, and each was controversial.

Today, Agassiz's logic seems unassailable, and no one doubts the reality of ice ages. To present-day scientists, as to Agassiz, the conclusion that there were great sheets of ice covering the northern continents is a straightforward outcome of the observations. However, at the time it was a radical concept. By the time Agassiz's book was published, many scientists had come to accept that the Alpine glaciers had once been somewhat more extensive than they were in the 1830s—after all, there were historical accounts as proof. Europe was just then emerging from a period of several hundred years of cool temperatures that would later come to be known as the "Little Ice Age." The slightly larger Alpine glaciers described in historical accounts could explain some of the geological observations in valleys now free of ice. But an ice age that was global in extent—that was a different matter altogether.

Agassiz's interest in glaciation may have been stimulated in part by his biological interests. Through his work on fossil fish, he was well aware of the paleontological evidence for periods of massive extinction in the Earth's past. The ice age theory provided a way to understand at least one of these events, and the frozen mammoths confirmed the extreme biological impact. In a passage from his book that is quite poetic, he describes the ice age landscape and the effects he believed the frigid climate must have had on living things:

> The development of these huge ice sheets must have led to the destruction of all organic life at the Earth's surface. The land of Europe, previously covered with tropical vegetation and inhabited by herds of great elephants, enormous hippopotami, and gigantic carnivora, was suddenly buried under a vast expanse of ice, covering plains, lakes, seas, and plateaus alike. The movement of a powerful creation was supplanted by the silence of death. Springs dried up, streams ceased to flow, and the rays of the sun, rising over this frozen shore (if they reached it at all) were greeted only by the whistling of the northern wind and the rumbling of crevasses opening up across the surface of the huge ocean of ice.

Like many of his contemporaries, Agassiz equated life with "a powerful creation." Religion still had a strong influence on thinking about the origin and history of life, even if observation and reasoning had gradually overturned the teachings of the religious authorities on matters such as astronomy and even the Earth's history. It was to be a topic that dogged Agassiz throughout his career. In spite of his contributions to paleontology and the evolution of fish, he would never accept Darwin's ideas on evolution.

Agassiz's lecture at the Neuchâtel meeting and publication of *Études sur les glaciers* were the first major parries in the debate about continental-scale glaciation. Throughout, Agassiz never flagged in his efforts to convince others about the reality of an ice age. The debate raged on for much longer than he, or for that matter anyone else, could have predicted. He continued to spend summers studying Alpine ice, and in 1840, he and his colleagues set up a permanent camp and observatory on one of the major alpine glaciers in order to make continuous observations of temperature, ice movement, the nature of the rocky moraines that characterize glaciers, and many other features of these "rivers of ice." The little scientific encampment quickly became a magnet for visiting geologists and inquisitive travelers. As always, Agassiz needed money for his venture. His friend and patron from Paris days, Alexander von Humboldt, suggested to the king of Prussia that Agassiz's glacier research was a worthy cause, and funds soon appeared. With characteristic indifference to his finances, Agassiz spent the money almost instantly on supplies, salaries for assistants, and equipment. He seemed always to be in debt.

Summers may have been spent on the ice, but during the rest of the year, Agassiz found time to travel widely within Europe, mainly in connection with his work on fossil fish. He visited museums and private collections wherever they existed, usually with his faithful artist assistant, who would make detailed sketches. His reputation grew, and he was rewarded with grants and honors, and also with long-lasting friendships with leading figures in geology and paleontology. Agassiz also used his travels to search for signs of glacial action on the landscape throughout Europe. Britain was considered to be the center for geological research at that time, and Agassiz, together with prominent British naturalists, made field excursions in Scotland, northern England, and Ireland. They found abundant evidence of past glacial activity, especially in the form of moraines, erratic boulders, and glacially scratched rock surfaces. These efforts won him some converts, but while many were willing to concede that the glaciers of the Alps had once been more extensive, it was much more difficult to persuade them that a great ice sheet had once covered much of the British Isles. For one thing, Britain is not a mountainous country, and glaciers were still inextricably linked to mountains in the minds of many geologists. For another, some large-scale landscape features that were being touted as glacial features, such as lakes and valleys, seemed just too vast to have been formed by the action of ice. It was really just a problem of imagination; most geologists simply could not imagine the scale of the long-gone continental ice sheets. They had been several kilometers thick and hundreds to thousands of kilometers in extent. Not a tree or a blade of grass survived where they stood; hills and valleys had been completely buried in a vast, monotonous white mantle of ice and snow. And they had moved, slowly but inexorably flowing under their own great mass. Agassiz once referred to glaciers as "God's great plough." He used the expression as a metaphor for natural catastrophe rather than a description of glacial erosion, but it is nevertheless an apt portrayal of the power of ice to shape the landscape, to scoop out lake basins, gouge out valleys, and pile up the loose debris in great mounds and ridges.

Agassiz was in his late thirties, at the peak of his scientific career and much appreciated in Switzerland, but in March of 1846, he sought greener pastures: he departed for the United States. Ostensibly, the trip was to give a series of invited lectures at the Lowell Institute in Boston, and also to lead an expedition, once again with funding from the king of Prussia, to explore the natural history of America. Agassiz spoke of returning to Neuchâtel to continue his work there, but few believed him. A poor manager, he was deep in debt because of an ill-conceived scientific publishing venture he had established in Neuchâtel so that he would have control over the publication of his monumental series of volumes on fossil fish. Partly because of his financial problems and partly due to the demands of his scientific work, his personal life was also in turmoil—his wife had moved back to her native Germany with two of their three children. And, although Agassiz was already recognized as one of the preeminent naturalists of his day, he had realized for some time that he would have to move on if he were to achieve all of his ambitions. The previous summer he had quietly turned over responsibility for his permanent glacial observatory to another man, one Daniel Dollfus-Ausset. He seemed to be tidying up loose ends. For many, his departure from Neuchâtel had the aspect of a permanent farewell.

The premonitions of his Neuchâtel friends and students were justified. Agassiz settled permanently in the United States, and only once returned, briefly, to the small town where he had accomplished so much of his best work. His departure from Switzerland also marked a turning point in his active work on glaciation. Although he lectured on the topic to great acclaim in the United States and made observations of glacial features both in North and South America, most of his time was occupied with work in zoology, and, increasingly, with administration and organizational activities. The days of slogging up Alpine icefields, measuring the slow, plastic flow of ice, or being lowered into a drain-hole that had been bored into a glacier by summer meltwater—he referred to this escapade as a "descent into hell"—were over.

This is not to say that Agassiz's personal interest in the subject ever flagged; it just didn't occupy the same place in his life that it had during the decade from 1836, when he was first convinced by de Charpentier and Venetz about the reality of past glaciation, to 1846, when he left for America. Indeed, one of Agassiz's first acts on reaching North America (after an Atlantic crossing so rough that rumors abounded in Europe that the ship, and Agassiz with it, had been lost) was to look for signs of glacial activity. En route to Boston, the ship had docked first at Halifax, Canada. Within minutes of stepping ashore, Agassiz had found what he was looking for on the hill overlooking the harbor: the same glacial grooves and scratches on the bedrock that he knew so well from the Alps. Such evidence bolstered his confidence. More than ever, he was sure that much of the northern hemisphere had once been covered by a deep ocean of ice. And in the coming years, he was to observe and describe signs of glaciation—moraines, glacial grooves and scratches, erratic boulders—throughout the northeastern United States. But as

time went on his observations became more cursory and his specula-
tions more grandiose. In 1865 and 1866, he conducted a long expedition
to Brazil. He reported seeing "glacial drift" and erratic boulders deep in
the Amazon basin, and within a week of returning to the United States,
he presented a paper to the National Academy of Sciences claiming that
large tracts of South America had been covered in ice during the ice
age. He provided very little firm evidence for this hypothesis. Many
geologists had already studied the deposits of the Amazon Basin without reporting any signs of past glaciation; Agassiz simply claimed that
most glacial features had been destroyed by the tropical climate. In his
enthusiasm for his idea, Agassiz was overreaching himself. He was
right about a global ice age, but he was wrong about the Amazon
Basin—in the Andes, and in the far south of South America, glaciers
had indeed encroached far beyond their present boundaries, but the
tropical lowlands of Brazil had not been glaciated.

By this time in his career, with other pressing responsibilities,
Agassiz was no longer active in science on a day-to-day basis. Younger
zoologists and biologists, both in Europe and in North America, were
exploring Darwin's ideas about evolution and generally supporting
them, but Agassiz continued to hold a catastrophist view of evolution,
partly a holdover from his early experiences working with his mentor,
Cuvier, in Paris. Colleagues and critics alike suspected that his claim
about the ice age having affected much of South America might be
influenced by his views on evolution—a truly global ice age would have
had more far more extensive and catastrophic biological consequences
than one that affected only some parts of the Earth. As we have already
seen, Cuvier had believed that evolution occurred through catastrophic
events that wiped out very large numbers of organisms simultaneously.
In his view, the new species that later arose were *completely* new, with
no connection to those that had gone before. Agassiz held similar
beliefs. We now know that the ice age did have significant biological
effects and caused species extinction even far from the ice-covered
regions. But these effects were much different from the catastrophism

Figure 3.   Louis Agassiz late in his career, as a professor at Harvard University. Agassiz cut an imposing figure and was a superb teacher. He changed the approach to teaching science in the United States by insisting that students "learn by doing." Photograph courtesy Ernst Mayr Library of the Museum of Comparative Zoology, Harvard University. Copyright President and Fellows of Harvard College.

espoused by Agassiz; they included environmental stress, loss of accustomed habitat, and the various environmental effects of climate change. Although the rate of extinction increased during the ice age, many species survived by adapting or migrating to more favorable regions.

In many respects, the later years of Agassiz's career are a paradox and a disappointment—the man whose curiosity, superb observational abilities, and penchant for synthesis led to great advances in zoology and geology early in his career gradually became stubborn and dogmatic as the years passed. Many of the implications of his work in biology, especially as they impacted ideas about evolution, were left for others to work out.

Although Agassiz's scientific work in the United States never did quite match the achievements of his early career, he nevertheless left another sort of legacy. When he arrived in Boston, Agassiz promptly charmed the influential citizens of that city and, as he had done upon taking up his position in Neuchâtel, set about popularizing science through his lectures. They were hugely popular, and he became a sought-after speaker. Not only was he a working scientist and expert who was anxious to explain his ideas to the world, but he was also outgoing and, with his Swiss accent, slightly exotic. His enthusiasm was contagious and his talks about ice ages caught the imagination of the public. Agassiz was so successful as a speaker that—appropriately enough for a newly arrived American—he was able to retire a considerable part of the debt he had accumulated in Switzerland from his speaker's fees.

Before long he was also a familiar figure in the young capital, Washington. He became a professor at Harvard (figure 3) and founded the Museum of Comparative Zoology there, and he changed the way science was taught by insisting that his students do hands-on work in the laboratory and the field. He was instrumental in founding Cornell University, the National Academy of Sciences, and the American Association for the Advancement of Science. Longfellow wrote a poem for him on his fiftieth birthday, and Oliver Wendell Holmes wrote another on the occasion of his departure on his expedition to Brazil. He published eagerly awaited articles on natural history in the *Atlantic Monthly*. When he died, in 1873, the vice president of the United States, the governor of Massachusetts, and many other notables attended his funeral.

Fittingly, Agassiz's grave at the Mt. Auburn cemetery in Cambridge, Massachusetts, is marked with a large granite boulder, retrieved with considerable difficulty from a moraine of the Aar glacier in Switzerland, near the spot where he had set up his glaciological observatory in 1840. There was another memorial to Agassiz as well. By the time he died, the reality of ice ages was recognized by scientists around the world.

CHAPTER FOUR

The Evidence

What, exactly, are the clues that betray the presence of extensive continental ice sheets in our planet's recent past? Some have already been described in previous chapters, and if you live north of about 40 degrees latitude in North America, or a bit further north than that in Europe, or in a mountainous region almost anywhere, you have probably seen some of the effects of glaciers for yourself—although you may not have realized it. Today many more features are recognized as having originated in the glacial-interglacial cycles of the Pleistocene Ice Age than was the case in Agassiz's day—everything from dead coral reefs in Indonesia now on dry land well above sea level to the rich soils of the central United States, developed on wind-blown silt called loess. The biosphere—the world of living things—was also strongly affected, although not to the degree that Agassiz thought. Careful examination and analysis of glacial effects, especially over the past few decades, has provided a remarkably detailed picture of how our planet has operated during the current ice age. As we shall see later in this book, there is even good (but circumstantial) evidence that the development of modern humans was influenced by the fluctuating climate of the glacial cycles.

In the decades after the concept of an ice age was first introduced, much of the debate about its validity centered around interpretation of purported glacial features in places like Scandinavia, Scotland, or the northern United States, far from the Alpine glaciers that Agassiz had studied. Opponents of the ice age theory searched for alternative explanations, but for those who were convinced early on that an ice age had occurred, this was a period of intense exploration and discovery. By the late 1800s, there were credible glacial maps that showed the former extent of ice in Europe and North America. The makers of those maps also quickly came to the conclusion that there had been a series of ice advances and retreats, rather than a single period in which glaciers grew to a maximum extent and then declined.

Both the initial proposal that there had been an ice age and the subsequent discovery that there had been multiple ice advances and retreats rested heavily on the significance of two types of glacial deposits: one that we have already encountered, erratic boulders, often referred to simply as erratics, and a second, glacial drift—a general term for the loose, rocky debris distributed across the countryside by glaciers. Erratic boulders are spread over large regions in Europe and North America; the most remarkable ones are very large and they are often quite different in makeup from the local bedrock. Large granite boulders like the one in figure 4 sit enigmatically on the local limestone in the Jura Mountains of Switzerland and western France, not far from where Louis Agassiz was born. How they got there was unknown before the glacial theory was developed—the nearest outcrops of granite are a hundred kilometers or more away. Some of them are so massive that they are difficult or impossible to move, and farmers clearing their land left them where they lay—great rocky sentinels sitting mutely in the middle of fertile fields, dispassionately surveying their surroundings. Similar features are common in the farmlands of the northern United States and Canada. In the northeastern United States, where there are outcrops of very distinctive rock varieties, trails of erratics of a specific type can often be traced for hundreds of kilometers, fanning out over the countryside "downstream" of their sources. Careful mapping of these erratic trails can provide an accurate picture of how the glaciers that carried the boulders moved across the land.

Figure 4. A large erratic boulder in a field near Örebro, Sweden. Boulders like this one, very different in makeup from other rocks in the vicinity and much too heavy to have been carried by water, convinced Louis Agassiz that large tracts of Europe had once been covered by thick, flowing ice that carried the boulders far from their place of origin. Note the woman to the right of the erratic for scale. A boulder of this size probably weighs close to ten thousand metric tons. Photograph copyright Dr. John Shelton.

Before the ice age theory was generally accepted, most geologists and naturalists argued that the erratics had been transported by water to their current resting places. They realized that even fairly small boulders would sink instantly in normal streams, but they also understood enough about rare natural phenomena such as tsunamis ("tidal waves") and great storms to know that water could transport heavy objects under extreme conditions. In the late eighteenth and early nineteenth centuries, memory of the great Lisbon earthquake of 1755 was strong. It had actually occurred off the coast of Portugal, not in Lisbon itself, but it had generated large tsunamis that scattered heavy objects far inland as though they were matchsticks. It was also well known that a raging mountain stream could carry very large rocks, especially when swollen with the downpour of a violent storm. But even such extreme events could not easily explain the massive granite erratics in the Jura Mountains, especially the ones that were perched high on valley walls, far above the streams below. Nor could they account for the presence in northern England of erratic boulders that appeared (based on their mineral makeup) to have originated across the North Sea in Norway, or those in the German lowlands that were hundreds of kilometers from their source. Compounding the problem of interpreting these deposits, however, was the fact that at a few localities in Britain, where much of the most detailed research into the ice age controversy was being conducted, the fine-grained drift that accompanied the erratic boulders contained seashells. Critics of the ice transport hypothesis seized on this; they claimed it was conclusive evidence that the ocean was involved. They argued that the erratics must have been transported by great, violent floods coursing over the land, and they said that there were simply no modern-day counterparts. They knew that the sea had covered parts of the continents in the past, because fossilized fish were found throughout Europe. The marine shells in "glacial" drift, they asserted, were proof that the sea had invaded the land yet again and left the drift behind when it receded. Actually, until about the 1820s, there was widespread belief that *all* of the loose sand and boulders strewn across the land surface had been deposited there by one or more floods, probably by the one described in the Bible. It was not until much later that the true origin of the seashells in drift was realized. James Croll, a Scottish scientist whom we shall encounter later in this book, deduced that they too had been transported by glaciers, scraped up by the ice along with sediments from the shallow seas around Britain and carried inland. However, before their origin was understood, the shells were a serious difficulty for those who argued that drift and erratics were ice age deposits.

Still, notwithstanding the seashell argument, even some of the opponents of the glacial theory had to admit that it would be difficult, if not impossible, to transport large erratic boulders in water over long distances, no matter how violent the storm or flood. It could be shown by simple physics that it couldn't be done. So they came up with the ingenious solution mentioned in chapter 2: the erratics might indeed have been carried by ice, but ice that was floating on formerly more extensive seas, transporting boulders from a northerly source. If parts of the continents had been submerged in the past, they reasoned, the icebergs could have floated over the sunken land, dropping their rocky burden as they melted. That would explain the presence of ocean shells in the drift. It was the idea of drifting icebergs that first led to use of the term "drift" for the characteristically chaotic sediments left behind by glaciers—sediments that have neither the well-defined layers nor the uniformity of grain sizes that characterize those deposited in water. The term is still in use today. Geologists also refer to such material as being unsorted, because it encompasses materials ranging in size from grains of sand and occasional shells to the erratic boulders themselves.

Drift and erratic boulders were not the only glacial features studied by early ice age researchers, but because essentially identical materials could be observed directly associated with glaciers in the Alps, these deposits were among the most persuasive evidence of past glaciation. Every existing mountain glacier carries a large amount of rock debris that will eventually become glacial drift. Some of it falls onto the glacier surface from the surrounding valley walls, and some is actually plucked from the bedrock below by the ice itself. Beginning with Agassiz's systematic studies at his glacier observatory, a series of investigations also showed that glaciers flow, and do so at significant rates. The rock debris is carried along with the flowing ice, and, at the snout of the glacier, dumped in a chaotic pile of large and small boulders, gravel, sand, and silt—a feature known as a moraine. Actually, glaciologists distinguish many types of moraines, but in its most general sense, the term—like the term "glacial drift"—just refers to the debris carried by a glacier. Terminal moraines mark the farthest extent of a glacier, lateral moraines form along the sides of mountain glaciers, and medial moraines in their middles, the result of tributary glaciers entering the main ice

Figure 5. That glaciers flow is particularly apparent from the air. This glacier in Greenland flows toward the observer, carrying on its surface ribbons of rocky material that have fallen onto its surface from the valley walls, forming moraines. Because many tributaries join the main glacier, it becomes more and more banded with moraines downstream as each tributary adds its contribution of debris. Photograph courtesy Professor Michael Hambrey, Liverpool John Moores University.

flow. Figures 5 and 6 illustrate a few varieties of moraines. Some types—for example, medial moraines—may exist on the ice of an active glacier, but can also be distinguished long after the glacier has melted away, because they form a longitudinal ridge in the middle of a glacial valley.

Ice in a glacier, like water in a stream, flows under the influence of gravity. A mountain glacier accumulates snow at its upper end, where the average temperature is low, and loses ice by melting at its snout. In places like Greenland and Alaska, some glaciers flow directly into the sea, where gigantic chunks break off and form icebergs. If snow accumulation and melting are more or less in balance over a significant

Figure 6. A cross section through a lateral moraine in Switzerland illustrates the great range of sizes of material it contains. Someone has sorted out piles of sand and boulders of various sizes near the bottom of the picture. Notice that a forest and a layer of soil has developed on the moraine. Glacial deposits from each cycle of the Pleistocene Ice Age have such soils developed on them, indicating that the ice advances were separated by long and relatively warm interglacial periods. This particular moraine can be traced for many kilometers. Photograph copyright Dr. John Shelton.

length of time, the size of a glacier and the location of its lower end will remain approximately constant, in spite of the fact that the ice is flowing and transporting rock debris all the while. When this occurs, the glacier is in a steady state, and very large terminal moraines can be built up. If the climate warms and the glacier melts away, the moraine remains as a distinctive landform—a great ridge composed of pebbles and boulders, marking the previous terminus of the glacier. Sometimes there is a whole series of these features, tracing out positions where the glacier front remained stationary for varying lengths of time before melting back further.

Once the nature of moraines formed by contemporary glaciers was understood, it became clear that much of the enigmatic "drift" so common in northern Europe and North America must have an analogous origin in the now-vanished glaciers of the ice age. The moraines left behind by continental-scale ice sheets are really not much different from those of alpine glaciers, except that they exist on a much grander scale. In places they can be traced for hundreds of kilometers, winding through the countryside and marking an ancient glacial boundary. But the ice sheets of the Pleistocene Ice Age didn't deposit their rubbly burden only as terminal moraines, easily recognized by their ridgelike shape. Some of that material was simply scattered across the landscape as a layer of gravel and boulders without any particular form. Sometimes the drift was shaped by the moving ice into features such as strange teardrop-shaped hills called drumlins, which usually occur in swarms, lined up parallel with one another. Exactly how drumlins form is unclear, but they apparently take shape beneath the flowing ice, their orientation reflecting the direction of ice movement. In other places, drift occurs as long, sinuous ridges of sand and gravel called eskers, which have occasionally been put to use as beds for railway lines in low-lying marshy areas. Eskers are thought to be essentially "negative streams"—rocky material built up in a confined stream that flowed beneath a glacier. When the glacier finally melted away, they were left standing above the surrounding countryside.

Starting soon after Agassiz published his *Études sur les glaciers,* geologists began to map out these features wherever they existed. A primary goal of this mapping was to determine the extent of the ice age glaciers, another to discover how they had flowed. Even now, details are being added to the general picture, which emerged quite quickly. It has become clear that the ice age glaciers did not form a single, gigantic ice sheet that extended southward from the North Pole, as Agassiz and his supporters had initially assumed. Instead, there were centers of ice accumulation, located where temperatures were low and the snow supply was ample. In North America alone, there were several centers of thick ice accumulation, with ice flowing out in all directions and in places coalescing with the glaciers of other centers. But some parts of the far north—for example, parts of Alaska—had no glaciers at all, even during the coldest part of the ice age, because of low snowfall.

It was the mapping that revealed the multiple glacial episodes of the Pleistocene Ice Age. There is an exceedingly simple but very powerful concept in geology, first formalized in the 1600s and still taught to beginning students in the earth sciences: any geological feature that cuts into or across another is younger than the one it cuts across, and any material deposited on top of something else is younger than the underlying material. To beginning geologists, it often seems silly to formalize such a commonsense principle, yet even quite complex sequences of geological events can often be unraveled by applying this concept. It has been used for everything from exploration for oil to working out the cratering history of the moon. When it was applied to the moraines and other deposits left by glaciers of the Pleistocene Ice Age, it showed that there had been several distinct glacial episodes, separated from one another by significant amounts of time.

The principle of superposition, as the concept just described is sometimes called, provides information about relative time—one deposit is older than another, or some process occurred before another—but not absolute time in years. That only became possible more than a century after the ice age theory was proposed, after the discovery of radioactivity and the development of techniques that used radioactivity for dating. But even in the nineteenth century, geologists were able to determine that there had been at least three and perhaps as many as five separate expansions of ice far south into Europe and North America during the Pleistocene Ice Age, and that these had been separated by long periods of time with much warmer climates. European and North American scientists gave these episodes different names, and it was not possible to correlate them precisely between continents; however, it was generally agreed that on each continent, the glacial deposits recorded the same series of cold and warm episodes. The changes in ice age climate had been global, or at least they had affected widely separated parts of the Northern Hemisphere similarly. We now know that the glacial periods identified by mapping their deposits were only the last few of a long string of cold and warm cycles stretching back several million years. This knowledge comes not from studies on land, but rather from evidence of a quite different type contained in deep-sea sediment cores. On land, the evidence for the earlier glacial cycles has been almost completely obliterated by the more recent ones, but in the oceans each layer of sediment buries and preserves the ones that preceded it.

How did the early investigators, without the help of radioactive dating methods, conclude that long time intervals separated the glacial periods? It was a task that required a certain amount of ingenuity. In many localities, it was fairly straightforward to use the principle of superposition to determine that there had been several different glacial advances. In places, younger drift could be observed deposited on top of earlier glacier debris, and in other localities, older moraines had been broken through and partly scoured away by more recent glaciers, which had deposited their own debris far beyond. However, determining just how much time had elapsed between these various events was a difficult problem. An important clue was that between successive glacial advances, soil had developed on the moraines and drift deposits. Soil forms anywhere rocks are exposed to rain—water is an effective solvent, and it also promotes chemical reactions with the minerals contained in rocks. The result is that solid rock dissolves and crumbles and is transformed into the soft clay of soil. Plants, insects, and microbes appear, churning the soil, facilitating even more chemical reactions and adding organic matter. In tropical climates with heavy rainfall, soils have been observed to form on fresh volcanic lava flows within a few hundred years or even less. But in the colder regions from which the ice age glaciers retreated, soils formed much more slowly. Soil layers that developed on moraines and drift between glacial advances indicate that the cycles were separated by relatively long periods of moderate climate. Fossils of plants and animals in the soil paint a similar picture. The interglacial periods were long enough for there to be a complete change of fauna and flora, and the new species were characteristic of temperate rather than arctic regions. When radioactive dating methods became available, it was discovered that through the last six or seven glacial cycles, the times of maximum ice advance were separated by roughly one hundred thousand years, and the warmer periods, with temperatures similar to today's, lasted ten to twenty thousand years.

When temperatures rise above freezing, large amounts of meltwater flow across the surfaces of glaciers, along their edges, at their bases, and even through the ice itself. During one of his field sessions on an Alpine glacier, the ever-curious Louis Agassiz had himself lowered down an almost vertical tunnel that had been cut by summer meltwater. It was one of his more foolhardy experiments; the tunnel got narrower and narrower, and eventually bifurcated, and Agassiz lost voice contact with his colleagues on the surface. They kept lowering him, right into an icy torrent deep in the glacier. Fortunately for Agassiz, the glacier-bound stream wasn't very deep, and eventually his friends became concerned and hauled him up. But it could have been much worse—the amounts of water coursing through melting glaciers can sometimes be huge.

The meltwater flowing away from a glacier carries with it grains and fragments of rock that were originally embedded in the ice. Ice is indiscriminate about what it carries, but the running water is quite efficient at sorting out the chaotic jumble of particles according to size and weight. It winnows the unsorted glacial drift, dumping the largest pieces at the base of the glacier or close to its boundary, and carrying the smallest grains in suspension over long distances. The meltwater streams build up sand bars in some places, gravel bars in others, and when they are flowing at full force, they sometimes carry quite large boulders along with them. In the northern United States and in Canada, in Scandinavia and northern Europe, man has taken advantage of this combined production and sorting process that is a relic of past glacial action. The sand and gravel deposits of the meltwater streams are scooped up by the truckload and used as construction materials. In overall economic importance, these deposits overshadow all other kinds of mining activity.

In these same regions, the action of ice and meltwater has left an aesthetic legacy in addition to a practical one: the undulating topography (and the sand traps) of many a well-groomed golf course.

The very finest particles of rock carried by the meltwater streams, much smaller than sand grains, are sometimes called rock flour. They are produced by the scouring action of the ice on underlying bedrock, and they are so fine that they remain suspended for very long periods of time and give glacial lakes their characteristic brilliant blue-green color. The scraping and scratching and polishing that produces rock flour leaves very distinctive telltale marks on the underlying bedrock. But it is not the ice itself that does the grinding; even hard, brittle ice at temperatures well below freezing cannot gouge out scratches and grooves in solid rock—it is simply not hard enough. Yet Agassiz and other careful observers of the Swiss glaciers found such features on rocks near the edges of glaciers. They noticed that the scratches were most prominent in areas where the ice had recently retreated, but, like the erratics and drift, could also be found far from any contemporary glaciers. They soon realized that it was actually the rock debris carried by the ice that was producing the scratches. Rocks and pebbles embedded in the ice were being dragged across the underlying surface; the glaciers were like gigantic sheets of sandpaper smoothing out the rocks beneath. In the process they produced the rock equivalent of sawdust: glacial rock flour. When scientists were eventually able to map out the movement of ice within glaciers, they discovered that the base of a glacier is continually being renewed with ice from above, complete with its embedded grit. The natural sandpaper is constantly being refreshed.

In most places that were glaciated during the Pleistocene Ice Age, scratches and grooves and polished rock surfaces are very abundant. Once you know what to look for, they seem to pop up everywhere. Recently, I walked along the wonderful meandering stone wall built by the artist Andy Goldsworthy at Storm King, an art park not far from New York City. The wall was made from local stones, with those in the top layer chosen for their flat surfaces. It was the Pleistocene glaciers that left them with these surfaces—most show the characteristic scores

Figure 7. A cartoon sketch of Professor William Buckland by the mining engineer Thomas Sopwith titled "Costume of the Glaciers," showing Buckland dressed for fieldwork. The numerous captions are difficult to read, even in the original, but the lines at Buckland's feet are noted to be "Prodigious Glacial Scratches" produced by "the motions of an IMMENSE BODY." Other captions are summarized in the text. From Mrs. E. O. Gordon, *The Life and Correspondence of William Buckland* (London: John Murray, 1894).

and scratches of glacial scouring. They are a reminder that just twenty thousand years ago the region was thick with ice.

In Britain, when the debate over the ice age theory was raging in the middle of the nineteenth century, a well-known mining engineer named Thomas Sopwith gently poked fun at the subject of glacial scratches. He sketched the Reverend William Buckland, professor of geology at Oxford, equipped for a field expedition, nattily attired in long coat and high boots, with scratched and grooved rocks at his feet (see figure 7). A caption indicates that one set of scratches had been produced by glaciers thousands of years ago. Another set, the label says, had been made by the wheel of a passing cart "the day before yesterday." Sopwith signed his cartoon with the words "Scratched by T. Sopwith." Buckland was something of an eccentric, a larger-than-life figure, well known to the public and also a highly respected scientist. He was impressed by his friend Agassiz's work on fossil fish, but initially unconvinced about ice ages. However, after Agassiz personally showed him glacial features in the Alps, and a few years later accompanied him to study moraines, erratics, and glacial scratches in Scotland and England, Buckland was won over and became one of Agassiz's allies in the ongoing controversy. The public, too, was following the debate—and perhaps chuckling over the seriousness with which learned men treated these little scratches on the rocks.

The scratching and grinding that produces these marks is also the process ultimately responsible for a peculiar type of sedimentary deposit that was not known in Agassiz's day to be related to glaciers, but which is now recognized as a key indicator of past ice ages. Characteristically, the sediments of glacial lakes contain what are called varves, a term derived from a Swedish word for periodic repetition. The distinguishing feature of these sediments is a repeated pattern of layers. In detail, in a typical case, the layers are fairly uniform in thickness and sharply separated from one another. They consist of a lower layer of relatively coarse, silty material, overlain by a layer of much finer-grained particles, and then the pattern repeats itself. Each coarse-fine pair constitutes a single varve, which may be a few millimeters to a few centimeters thick. It is now known that each varve represents one year of sediment accumulation. There are some rare circumstances in which similar deposits can be formed in nonglaciated areas, but the vast majority of varves are the product of deposition in a glacial lake. In summer, when meltwater is abundant, streams carry the products of glacial scratching and grinding into the lake, where the coarsest particles sink to the bottom fairly quickly. The finest material—the rock flour—mostly remains suspended, in part because winds keep the water stirred up in summer, and in part because the particles are so small that they settle only very slowly. As winter approaches and the temperature drops, the streams of meltwater coming from the glacier gradually dwindle and eventually stop. There is no longer a supply of new sediment, but over the winter, with a frozen surface to keep the water still, the fine suspended particles slowly sink to the bottom to form the fine-grained part of a varve couplet. A giveaway that the varves are indeed of glacial origin is the presence of dropstones, occasional large pebbles or rocks that are embedded in the fine-grained varves. A bit like erratic boulders on land, they seem, because of their size, to have nothing to do with the varved sediments in which they are found. They are, in fact, quite literally dropped in. Carried out into the glacial lake on pieces of ice, they fall into the fine-grained sediments when their transport melts. If the sediments of a glacial lake harden into solid rock, they can survive long after moraines, drift, and other surface deposits have vanished. Often, preserved varves are among the few remaining indicators of past ice ages, as is the case for one of the Earth's earliest, dating from about 2.2 billion years ago. Varved sediments containing dropstones still survive from this episode at several places around the globe.

Even during the depths of the last glacial maximum of the Pleistocene Ice Age, summer melting occurred along the southern margins of the Northern Hemisphere ice sheets, generating streams, rivers, ponds, and small lakes with varved sediments. Vast regions along these ice margins were similar to the present-day arctic tundra, with little vegetation and huge amounts of rock debris. Some of the moraines and drift from those times remain intact—they are the same features that have been mapped to determine the extent of the ice sheets—but large amounts of debris were also carried away from the glaciers by meltwater streams, which sorted and winnowed the grains and pebbles. During dry periods, the smallest of these grains were also moved around by the wind, sometimes over very great distances. Much of the central United States, from the Rocky Mountains east to Ohio and Pennsylvania, is still blanketed by such wind-blown dust of glacial origin. This material is called loess, from the German word for loose, and in the Great Plains and elsewhere, it forms rich soils. Similar deposits, although smaller in extent, occur across Europe. In Asia, too, there are thick loess accumulations, and in a few places, caves have been excavated in loess cliffs to serve as dwellings.

Most of the North American and European loess can be traced directly to the outwash of glaciers. Not only does it occur roughly in a band along the southern margins of the former ice sheets, but its makeup matches that of the material in the moraines and drift that remain. However, the source of the thick accumulations of loess in China has been traced to the arid deserts of north China and central Asia, and the deposits seem to have nothing to do, directly, with glaciers. Nevertheless, they were formed during the Pleistocene Ice Age at the same time as loess of glacial origin was accumulating in other parts of the world. The Chinese loess deposits have become especially important for understanding the ice age climate, because they accumulated over many glacial-interglacial cycles. Drill cores through these deposits indicate that during the glacial intervals of the Pleistocene Ice Age, climates worldwide were cold, dry, and windy—hence the widespread transport of dust. During the interglacial periods, climates warmed, and it became wetter and less windy. Loess accumulation slowed down, or stopped altogether, and in many places, soils developed on the deposits. When temperatures dropped and the glaciers expanded again, the cycle was repeated. Most mathematical simulations of glacial climates predict more intense wind systems than today, especially close to the margins of the ice sheets. The Chinese loess indicates that the influence of ice sheets was felt even very far away from the main centers of ice accumulation.

It is fairly easy to envision how the fine-grained particles of loess and the coarser material in moraines and other glacial deposits are scraped and gouged out of the bedrock as grit-laden ice sheets move over them. It is less easy to imagine the cumulative effects of this glacial erosion on a large scale. The results can be quite spectacular. During glacial times, ice filled many valleys that are now occupied by quiet streams. The flowing glaciers effectively sandpapered these valleys on a grand scale, smoothing them out into a U shape in cross section, with steep walls and a flat bottom. To some extent glacial erosion also tends to straighten out minor twists and turns in the ice-filled valleys. In contrast, where glaciers have never been present, valleys typically have a V shape, the result of erosion by flowing water in a fairly narrow valley bottom. The beautiful, steep-walled fjords of Norway, British Columbia, and Chile all owe their existence to glacial erosion during the ice age. Yosemite Valley in California, a national park visited by millions of tourists, is another classic U-shaped valley gouged out by glaciers. Its spectacular waterfalls, cascading down the steep valley walls, are also products of glaciation. Glaciers, like streams, often have tributaries, but when the main glacier fills a valley, the tributaries can only join it at the level of the ice surface, far above the present valley floor. When the glaciers eventually melt away, the tributary valleys are literally left hanging— and the streams that run in them spill over the main valley walls, creating waterfalls. Like varves and drumlins and glacially polished bedrock, the features of glacier-cut valleys are unique and provide a long-lasting record of past glacial episodes.

Not all of the effects of the Pleistocene Ice Age are as obvious as fjords or glacial drift, however. One of the most important of these is something that is difficult to observe visually but is quite predictable if you think about it a bit. It is the fact that there was a massive transfer of water from the oceans to the land during the glacial periods. It has been estimated that at the maximum of the most recent glacial period, about twenty thousand years ago, sea level was approximately 120 meters lower than it is at present. Along most shorelines, dry land extended far out into what is now quite deep ocean water. A map of the world as it was then would look quite different from the one we are familiar with today.

One hundred and twenty meters over the entire ocean adds up to a very large amount of water, about 3 percent of the present ocean volume. All of this water, evaporated from the sea, was transported to the continents as water vapor in the atmosphere, fell as snow, and accumulated as the glaciers of the ice age. Where the ice was thick, as in parts of Canada and Scandinavia, for example, an enormous weight was placed on a relatively small area of the continental crust. Slowly but steadily, the crust actually sank down into the yielding rocks of the underlying mantle in response to this burden. When the ice melted, these same areas began to rebound, and they have been slowly rising for the past ten thousand years or more. Along some rising seacoasts, this uplift is chronicled by a series of "raised beaches," which look like bathtub rings, except that they record the uplift of the land, rather than the falling level of water.

Just how do we know that the accumulation of ice age glaciers lowered sea level by 120 meters? Well, it would be possible to calculate the amount of ocean surface lowering if you knew exactly how much ice was on the land, but that too is a difficult question to answer. You would need to estimate how thick the ice was, and what area of the land it covered. What seems a simple problem suddenly looks quite complicated, and to solve it required a great deal of ingenuity. Evidence began to accumulate when oceanographers, studying the nature of the seafloor close to the continents, found that many river channels continue under water, uninterrupted, far beyond the present-day shoreline. A classic example is the Hudson River, which has a deep channel extending far across the continental shelf. However, it is well known that rivers can no longer erode a valley once they enter the sea—instead, they typically deposit sediments and build up a delta. It was immediately clear that the now-submerged channels had once been above sea level. But exactly when this erosion occurred was not known, and the precise amount of sea-level lowering was also difficult to determine with certainty.

Enter Richard Fairbanks, a geochemist at Columbia University, who developed a program to recover drill cores from the coral reef platform surrounding the island of Barbados. The species of coral that Fairbanks was interested in grows only right at the sea surface, and as ice age glaciers melted and sea level rose, the coral had grown upward to keep pace. A piece of coral recovered from 50 meters down the drill core must have been at sea level when it grew, so by dating such a sample, Fairbanks could determine quite accurately the time when sea level was 50 meters lower than at present. Repeated analyses of this sort throughout the cores allowed him to plot the change in sea level over time. Twenty-thousand-year-old corals, he found, grew near an ocean surface that was almost 120 meters below present-day sea level. Fairbanks also found that the rate of sea-level rise since the time of maximum glaciation has been quite variable. Bursts of rapid increase alternated with periods of slower change, reflecting fluctuations in the amount of melting of the ice sheet, probably the result of irregular warming of the climate over the past twenty thousand years.

The corals that Fairbanks analyzed are currently under water, but there are a few places in the world where now-dead coral reefs are found on dry land, well above the present sea level. These corals, too, grew very close to sea level. Many have been dated to near 120,000 years ago, and this age is believed to mark the time of the last interglacial episode, the most recent time in the Pleistocene Ice Age when the climate—and the sea level—was similar to today's. Because land is not fixed—geological forces can cause it to move vertically independent of sea level change—there is some debate about exactly how high the sea level was during the 120,000-year-ago interglacial period. But it seems to have been at least as high as today, suggesting that something close to the present-day ice volume was the minimum reached during that warm period, before the ice began to advance again. Even if the exact ice volume is uncertain, one of the most important results from studies of these now-stranded corals is the accurate dating of the last interglacial period.

The rising and falling sea levels of the Pleistocene Ice Age have had some interesting and perhaps unexpected consequences. Consider Alaska about fifteen thousand years ago. The maximum cold of the glacial cycle had passed, but the climate was still dismal. However, unlike much of North America, the lowlands of Alaska were free of thick glaciers, because there was not enough precipitation to sustain an ice cap. Small bands of nomadic hunters roamed the land, following the big game that thrived in the damp, cold tundra. Woolly mammoths were their prize target, and the hunters often tracked them over long distances before attempting a kill. It was dangerous business—the beasts had thick hair, thick skin, and a very thick and protective layer of fat. The hunters, armed only with primitive weapons, were no real match for an angry mammoth. Many were probably killed or severely injured in the close encounters that were necessary to slay one of these gigantic animals. But the rewards were great when one was brought down. A single mammoth could feed, clothe, and supply a band for a long time.

The hunters had followed the mammoths and other large animals eastward from Asia across what is now the Bering Sea. Some of them may have traveled by small boat along the coast, but many walked. Twenty thousand years ago, at the height of the last glacial period, sea level was so low that dry land joined what are now separate continents. Slowly, imperceptibly, and probably unconsciously, hunters had moved across the land bridge and become the first immigrants to the new land. Without the ice age, North America might have remained unpopulated for thousands of years more.

It is obvious that the Pleistocene Ice Age has affected the Earth in a multitude of ways, some quite apparent and others much more subtle. Only a few of the more important ones have been touched on in this chapter. Once scientists had accepted the reality of the ice age and the most obvious features of glaciation had been deciphered, they faced an important question: What causes ice ages?

CHAPTER FIVE

Searching for the Cause of Ice Ages

Curiously enough, there is no evidence that Louis Agassiz ever spent much time thinking about what caused the ice age that he proposed. He seems to have been much more interested in effects than in causes. The fact that there was no easily grasped mechanism to explain the Earth's sudden descent into frigidity was undoubtedly one of the reasons the ice age theory was resisted for so long, despite the field evidence. Agassiz was content to amass the various clues that supported his conclusions about widespread glaciation and to fit the theory neatly into his own thinking about the history of life on Earth. But there were others who *were* concerned about causes. They knew that in many places in Europe, there were both fossils of tropical plants and animals and evidence of glaciation. The ages of the fossils weren't known, but the principle of superposition suggested that they predated the ice age—the glacial deposits lay above them. Sometime before the ice age, Europe had been tropical—what had caused the radical change? (Actually, it was discovered later that the tropical period in Europe had occurred tens of millions of years before the Pleistocene Ice Age, and the climate change wasn't as abrupt as it appeared to these early workers.) The other problem faced by those trying to understand the cause of the ice age was evidence of multiple episodes of glaciation. A whole series of alternating cold and warm periods had occurred. They had to find a mechanism that could account for the Earth's repeated cycles of freezing and thawing.

The debate about a possible cause for Agassiz's ice age was intense in Britain, a hotbed of geological activity at that time, and especially in Scotland, where the evidence for glaciation was abundant and the geological scene was lively and populated by prominent scientists. By the 1860s, twenty-five years after the ice age had first been proposed, most geologists were convinced that the Earth had been through a frigid period, with multiple cycles of heat and cold. There were some holdouts who still thought the glacial deposits were best explained as debris that had fallen from drifting icebergs, but they were in the minority. There was no lack of theories about why it had happened, but none of them were very satisfactory. They included, for example, the idea that the Earth's pole of rotation had shifted, so that polar regions lay close to the equator and temperate regions close to the pole; that the Earth had passed alternately through cold and warm regions in its travels through space; that the sun is a variable star, its energy output changing over time; and that the distribution of land and sea had been different in the past, leading to glaciation when the continents were clustered near the poles. This latter idea is not so far-fetched, and it presaged the theory of continental drift. The continents *have* moved over the Earth's surface, because of plate tectonics. It is also believed by many scientists that one of the conditions necessary for extensive glaciation is to have continents at or near the poles. However, the movement of continents is very slow, much too slow to be the cause of the geologically short glacial-interglacial cycles of the Pleistocene Ice Age.

A credible answer to the ice age puzzle, the first to gain widespread support, came from an unlikely source: someone who was little known in the world of science, a man who in one sense was an amateur, although an amateur who taught himself enough physics, chemistry, and mathematics to make significant scientific contributions. His solution to the problem has since been modified in its details, but the basic premise is the same one that is in favor today. The man's name was James Croll (figure 8).

Figure 8.   James Croll, the self-taught Scotsman who
became a well-known figure when he proposed an astronomical theory of ice ages in the 1860s. Although his ideas fell into
disfavor, they were eventually revived and revised, and today
there is general agreement about the importance of orbital
parameters for climate change. Photograph from James
Campbell Irons, *Autobiographical Sketch of James Croll*
(London: Edward Stanford, 1896).

It is hard to imagine two men more different than Louis Agassiz and
James Croll. Agassiz was a well-educated middle-class European, outgoing, articulate, charming, and ambitious. Croll was a poor Scot, taci-
turn, with very little formal education. He tried various occupations; at
one point in his life, he was an insurance salesman, something he found

painful because of the personal interactions it required with potential clients. Unlike Agassiz, he was not naturally at ease with strangers. It is easy to imagine Agassiz as a successful insurance salesman, but Croll hated the job. Relatively late in his career, after he had become well known for his work on the causes of ice ages, he was persuaded to put together some autobiographical notes. His words in preface to these provide a glimpse of his character:

> I have been frequently urged during the past few years to draw up a statement of the principle incidents of my life. As this is a thing to which I have a strong aversion, I have hitherto declined. Induced mainly by the desire of my wife, I have at last agreed to comply with the wishes of my friends. Mrs. Croll will hurriedly jot down in pencil, to dictation, the facts as they occur to my mind. These jottings will probably never be reviewed or read over by me . . . it is a sort of work to which I am naturally ill adapted . . .

But in spite of their differences, Croll had a tremendous curiosity and an inner drive to learn and to explore knowledge that was in many ways at least the equal of Agassiz's. He had an ability to focus on a problem or a topic to the exclusion of almost everything else, and he was resolute in his pursuit of answers to questions he encountered in his reading. At the age of twenty-one, while running a tea shop, Croll acquired a copy of *Freedom of Will* by the American philosopher-theologian Jonathan Edwards (1703–58). He was so enamoured with the book that he read and reread it minutely in order to gain command of all of the author's arguments. He would spend an entire day on a single page. Most of his spare time for a period of more than a year was devoted to Edwards's book. To his customers, he was an odd character: a large, shy man with a laborer's frame and hands, running a tea shop but studiously reading philosophy whenever he had a spare moment. It was clear to all—even to Croll—that he was not cut out for business, and his tea shop was short-lived. But his ability to concentrate his intellect on nearly any subject he found interesting never wavered. This trait served him well in pursuit of a theory for the cause of ice ages.

Based on his reading, Croll suspected that the cycles of glacial and inter-glacial climates might have an astronomical cause. To investigate this possibility, he had to work out the mathematics of the Earth's orbit around the sun. His goal was to calculate how changes in the orbit would affect the amount of solar energy that is received by the Earth— if the changes were large enough, they would surely influence the cli-mate. In the 1800s, there were no computers or calculators available— all of the calculations had to be done, painstakingly, by hand.

Perhaps fittingly for a man whose name was to become linked with the ice ages, James Croll was born on a cold, snowy January night in 1821, near the end of a time that is now known as "The Little Ice Age" (about which more will be said in chapter 11). But weather aside, there was little to indicate that this new entry into the world would make important contributions to our understanding of ice ages and climate. Croll was born in rural Scotland; his father was a crofter who also worked as a stonemason. It was a not an easy life. Although the family was by no means destitute, neither were there any luxuries. Young Croll was somewhat sickly, and only attended formal school classes intermit-tently. He was educated partly at home, sometimes tutored by his par-ents and at others by a local schoolteacher. But at least initially, he showed little real interest in learning. However, when he began to real-ize the possibilities that education offered, his attitude changed. Unfortunately, just at that time, at age thirteen, when he was finally eager to push on with his learning, he was forced to abandon further study. He was needed to work on his family's small farm.

Croll's change of heart about education seems to have come from a single incident in 1832, when he was eleven. It was characteristic of a number of events in his life that suddenly led him in new directions, often, it seems, on little more than a whim or passing impulse. The 1832 incident occurred on a visit to the nearby city of Perth, where the young Croll found the very first issue of a small periodical called the *Penny Magazine* in a bookshop. He promptly bought it, and was hooked— from then on he purchased issues whenever possible. The little journal was published by an organization that called itself "The Society for the Diffusion of Useful Knowledge," and in the case of James Croll, it certainly succeeded in its aim. Although Victoria was not quite yet queen when Croll stumbled on that first issue, it was a magazine and a venture typical of Victorian Britain: eclectic, wide-ranging, and meant to bring information about the wider world to the British public. The issue that first caught young James's eye in Perth contained, among other things, brief sketches of the lives of the French mathematician and philosopher René Descartes and the British physician who discovered the true nature of blood circulation, William Harvey; an article about Van Diemen's land (now Tasmania); and a story about the grizzly bear that had just arrived at the London Zoo. Whether it was the illustrations or the titles that attracted Croll's attention is unknown, but from what we know about his studious habits in later life, it is likely that he read every word in the slim, eight-page issue. For a country boy, it was a window onto lands beyond his own and a source of insight into the accomplishments of prominent intellectuals. It was almost certainly one of the major inspirations for his own intellectual endeavors. As a grown man, Croll went to some lengths to purchase a few missing back issues of *Penny Magazine,* which was then no longer published, so that he would own a complete set.

Stimulated by what he learned in *Penny Magazine,* but still tied down to work on the farm, Croll eventually purchased a few key books on philosophy and science and set to work to educate himself. By the time he was sixteen, he was well versed in a wide range of subjects—remarkably so, considering that for much of this time, his days were occupied with physical labor. Croll's approach probably helped. He was to remark later that he was generally not interested in the small details of a problem, but instead wanted to understand the underlying principles. The details would fall into place if one knew the framework.

Croll eventually became a prominent figure in science and philosophy in Britain. It is difficult to know whether he would have been even more productive had he been able to attend university and live the life of a scholar. As it was, the path from a self-taught farm laborer of sixteen to recognition as a leading intellectual was long and tortuous.

The sixteen-year-old farm laborer reading books on philosophy in his spare time really wanted to attend university. But he had a very mediocre scholastic record and no formal training in Greek, Latin, or mathematics—important subjects for an aspiring university student in the nineteenth century. He also had no money. With characteristic logic and determination, although perhaps without pondering the consequences too carefully—another one of those new-direction-on-a-whim turning points in his life—Croll mulled over his situation for a few days and decided to become a millwright. It seemed to him to be an occupation that would fit his abilities perfectly. I know mechanics, he thought, therefore I'll become a millwright. What he hadn't considered was that his penchant for understanding principles rather than details meant that he knew mechanics from a theoretical point of view, but had little feeling for the practical side of machines. It must have been a rude awakening. But he persevered, served out an apprenticeship, and joined a local firm. "It was on the whole a rather rough life," he reflected in his understated way. The work took him around the countryside repairing mill machinery, sleeping in "on an average, three different beds a week." Bed is perhaps too luxurious a description for some of the sleeping arrangements Croll and his fellow mechanics endured; usually, they were put up in an unheated shed or barn. To add insult to injury, his employers were having a difficult time financially, and the employees were not always paid. After five or six years of slogging it out, he decided he had had enough and quit, resolving to try something else.

Croll was then twenty-one, unknown to anyone save a few farmers and millwrights in one small part of Scotland. He was still as unlikely a candidate to make major contributions to science as he had been at birth. It is all the more amazing, then, that when, in his sixties, Croll requested a small increment to his meager pension, a long list of luminaries from across the country petitioned to support his cause. The duke of Devonshire, the marquess of Salisbury, the duke of Buccleuch,

Alfred, Lord Tennyson, Thomas Huxley, Joseph Lister, various members of Parliament, and many, many others lent their names to his request. It was a startling testament to his emergence as an original and influential thinker, especially as he had never been a part of the stratum of society occupied by most of his supporters.

After deciding to give up his chosen profession as a millwright, Croll, for the next sixteen years or so, worked variously as a carpenter, a tea merchant, a self-employed electric battery maker, a hotel manager, an insurance salesman, and a writer for a newspaper. None of these positions lasted very long, and they required him to move about from place to place in Scotland, and, at one point, to Leicester in England. During most of this time, the itinerant Croll read voraciously and systematically, concentrating on his twin loves of philosophy and science. He was much interested in the subject of will and the question of the existence of God. During a spell of unemployment when he was about thirty-five, he organized his thoughts on the matter and wrote a book: *The Philosophy of Theism.* Croll was still an unknown to those who pondered such matters, and he published the book anonymously, probably on the assumption that it was better to be anonymous than to be an unknown writer. Although he didn't put his name on the book, he was otherwise not at all secretive about being its author. The year was 1857 and this was his first real venture into print. The subject matter wasn't exactly likely to guarantee commercial success, and Croll had difficulty finding a publisher who was willing to take a chance on the book. However, in the end, it got good reviews. Although only five hundred copies were printed, Croll even earned a small sum in royalties. But perhaps the most important effect of this book was that it brought Croll recognition as a significant intellectual. He would no longer feel the need to publish his work anonymously.

A few years after his book appeared, in 1859, Croll's meandering, frequently changing career finally settled into a semblance of normalcy when he took a position as caretaker at Anderson College, a private school and museum in Glasgow. No more hard manual labor, no more soliciting life insurance contracts from strangers, no more traveling about the country. Croll's position required a minimal amount of effort, physical or mental, and his salary, although small, was sufficient for his needs. His brother, who was physically disabled and lived with Croll and his wife, helped him with the work. The school had a well-stocked library, and the science section was extensive. Croll had plenty of free time; for a man who loved to read and think it was heaven. Almost forty, he was ready to begin making his mark in the field of science.

In his short autobiography, Croll recounts that when he took up his position at Anderson College, his primary intellectual interest was in philosophy and religion, the twin subjects of the book he had published a few years before. He had plans to expand his earlier exploration of these topics in a more systematic way. But instead he found the library at his new place of employment full of interesting works in science, and he decided to put his "metaphysics" aside for a while to investigate them. As a teenager, when he first began his program of self-education, he had been much taken with physics and mathematics. Through the new-found resources of the library at Anderson College, he wanted to reacquaint himself with these subjects, and to find out what was new in the world of science. He became especially interested in research into the nature of heat, electricity, and magnetism, and, quite amazingly, within a few years began to make original contributions to these subjects.

The list of Croll's scientific publications begins in 1861, just two years after he began his caretaker's job in Glasgow. His first effort was a paper on electricity in the respected *Philosophical Magazine*. Over the next twenty-five years, he contributed, on average, several articles annually to the scientific literature, many of them in the leading periodicals of the day. Although it was undoubtedly a time when a determined "outsider" could make more of an impact on the world of science than is possible in today's specialized world, Croll's record is nevertheless impressive. It is one that would bring credit to a full-time present-day academic.

Although Croll was retiring by nature, he had no hesitation in sending copies of his scientific papers to leading scientists throughout the country and corresponding with them about his work. He seemed secure in his scientific endeavors and confident in his own ability, in contrast to his feelings about some of his other ventures. Recognition of his work was rapid. In 1863, the prominent physicist John Tyndall, professor of natural philosophy at the Royal Institution in London, corresponded with Croll about some of the ideas expressed in his early publications. Tyndall urged him to continue sending papers, writing, "I have no doubt that anything you send me will interest me."

At about this time Archibald Geikie, one of Scotland's most prominent geologists, published a lengthy and detailed description of glacial features in the country, concluding that they must be the work of great ice sheets that had flowed across the land in the past. The work had been discussed in scientific circles in Scotland for several years before its formal publication, and, attuned as he was to current research in the physical sciences, Croll was undoubtedly well aware of it. Here was a challenging and important unsolved problem to which he could turn his attention. Croll thought initially that it would be a relatively straightforward task; "little did I suspect, at the time when I made this resolution [to investigate the causes of ice ages] that it would become a path so entangled that fully twenty years would elapse before I could get out of it," he would later say in his autobiography.

Croll attacked the ice age problem in the same way he approached all of his work in science: through first seeking to understand the governing principles. He described this method in the introduction to his book *Climate and Time,* which was published more than a decade after he first became involved in the ice age debate. Writing about what he termed "The Fundamental Problem of Geology," Croll notes:

> We may describe, arrange, and classify the effects as we may, but without a knowledge of the laws of the agent we can have no rational unity. We have not got the higher conception by which they can be *comprehended.* It is this relationship between the effects and the laws of the agent, a knowledge of which really constitutes a science. We might examine, arrange, and describe for a thousand years the effects

produced by heat, and still we should have no science of heat unless we had a knowledge of the laws of that agent. The effects would never be seen to be necessarily connected with anything known to us; we could not connect them with any rational principle from which they could be deduced *à priori [sic]*. The same remarks hold, of course, equally true of all sciences, in which the things to be considered stand in the relationship of cause and effect. Geology is no exception.

Croll's words could profitably be taken to heart by some scientists today. They show quite clearly why he was so successful as a scientist: he was not content to tabulate observations, or even, as Agassiz did, to synthesize them into a theory, without first trying to understand the underlying principles. In the case of glaciation and ice ages, he recognized that a major determinant of climate is the amount of heat energy the Earth receives from the sun—that had to be, in his terminology, one of the "agents" of climate. He also knew that the heat received from the sun varies because of the constantly changing orbit the Earth follows in its travels around the sun, and he made those variations the centerpiece of his investigation.

Croll was not the first to attempt to relate climate to the effects of the Earth's variable orbit, but he was the first to put the theory on a firm scientific footing. The idea had actually been discussed on and off for almost a century before Croll entered the debate. By the 1850s, a section on "Astronomical Causes of Fluctuations in Climate" was even included in Charles Lyell's widely read *Principles of Geology,* generally considered to be the first true textbook of geology, then in its ninth edition. In his book, Lyell summarized the evidence that had been accumulated up to that time, but he remained lukewarm about the possibility that astronomical influences were important. However, he did suggest that someone should carry out the laborious calculations that would be necessary to fully investigate this possibility.

Prior to Croll's work, a few scientists had looked into the mathematics of the Earth's orbital variations, but those who had thought about the implications for climate had concluded that the changes

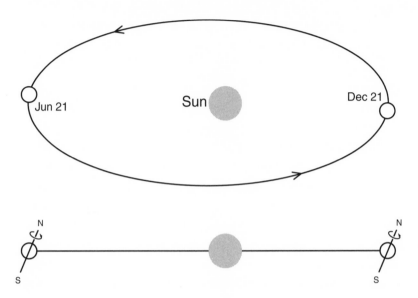

Figure 9. Plan (above) and side (below) views of the Earth's orbit around the sun. The eccentricity of the orbit is much exaggerated, and the sizes of Earth and sun are not to scale. As the eccentricity changes, the orbit becomes either more or less elliptical. The tilt of the Earth's axis of rotation is shown in the side view. At present, the Northern Hemisphere is tilted toward the sun in June, when the Earth is distant from the sun.

would not be of much consequence. They had considered the fact that the Earth's orbit around the sun is not circular, but is actually an ellipse, with the sun located at one of its two foci, slightly offset from its center (see figure 9). This means that as the Earth makes its yearly journey around the sun, it is sometimes closer (and thus receives more solar energy) and sometimes farther away. It was also known that the shape of the elliptical orbit changes over time, from being almost circular at some times to much more elliptical at others. Such changes are formalized in the concept of eccentricity—the more elliptical the orbit, the greater the eccentricity. Changes in eccentricity affect the amount of solar energy received by the Earth at various points in its orbit—when the orbit is almost circular, the energy received is almost constant throughout the year; when the orbit is more elliptical, the variation

between times of closest and farthest approach is much greater. But calculations showed that when averaged over a year—one complete revolution of the Earth around the sun—the differences resulting from
these variations are negligible. Few believed they could be responsible
for ice ages. There was an exception, however. In 1842, Joseph
Adhémar, a French scientist, wrote a book titled *Révolutions de la mer,*
in which he proposed that the combined effects of eccentricity and the
tilt of the Earth's axis of rotation play an important role in glaciation.

Because the details of the Earth's orbit are important for understanding the arguments made by Adhémar, Croll, and others about
glaciation, it is worth pausing briefly to consider them more carefully.
Without perturbing influences, the orbit of the Earth around the sun
would be unchanging. But due to gravity, each planet has an effect on
the orbits of the others. And because the planets orbit the sun on different timescales, their relative positions, and therefore their influences
on one another, are constantly changing. Thus the shape of the Earth's
elliptical orbit—its eccentricity—is also continuously changing. These
changes are slow and regular, and they can be calculated. Over long
periods of time, the Earth cycles through the same orbital conditions
over and over again. With computers and accurate knowledge of the
masses of the planets, it's possible to plot out the Earth's orbit, and
those of the other planets, very accurately far into the past or the
future. These kinds of calculations are routine for NASA engineers—
when they launch an exploratory spacecraft to another planet, they
have to know precisely where the Earth is in relation to other bodies in
the solar system and where the target planet will be several years later
when the spacecraft reaches it. The very first calculations of the planetary orbits, a truly monumental achievement in applying Newton's
gravitational theory to the solar system, were made by the French
mathematician and astronomer Pierre-Simon Laplace in 1773. By the
time Croll tackled the problem, the elliptical shape of the Earth's orbit
was well known, and it was also known that the eccentricity gradually
changes. However, no one had yet done the systematic calculations to

determine exactly what these changes were over very long periods of time.

In addition to having an elliptical orbit around the sun, the Earth exhibits a peculiar feature—it rotates around an axis (an imaginary line drawn through the north and south poles) that is tilted relative to the plane of its orbit around the sun (this is also shown in figures 9 and 10). The tilt today is 23 1/2°, but just why the axis is tilted is still a mystery. Some scientists believe that it is residual from a gigantic collision early in the Earth's history, when a small planet, about the size of Mars, crashed into the Earth and knocked it into its tilted position. Regardless of its origin, however, we're fortunate to live on a tilted planet—it's the reason we have seasons. You can work this out from the diagrams showing the Earth's orbit, or by experimenting with a flashlight and a round object to represent the tilted Earth. If the axis were perpendicular to the plane of the orbit around the sun, the length of day would be twelve hours everywhere, throughout the year. With a tilted axis, it changes.

As the Earth makes its yearly journey around the sun, the direction of tilt remains constant, so that at one point along the orbit, the North Pole tilts directly toward the sun (the Northern Hemisphere summer solstice) and at another it tilts directly away (the winter solstice). This simple picture is complicated, however, by the fact that the Earth not only rotates, it also wobbles, exactly like a spinning top (see figure 10). The wobble, caused by the combined force of gravity from the moon and the sun, is very slow on a human timescale, so that we don't notice it at all. But over time, the orientation of the Earth's axis of rotation changes, tracing out a circle that takes approximately twenty-six thousand years to complete. Today, the north end of the rotation axis points toward the North Star. Thousands of years from now, because of the Earth's wobble, it will point at a different part of the heavens, and some other star will have to be identified as the "pole star."

One consequence of the Earth's wobble is the phenomenon called the precession of the equinoxes. Because of the 26,000-year-long wobble

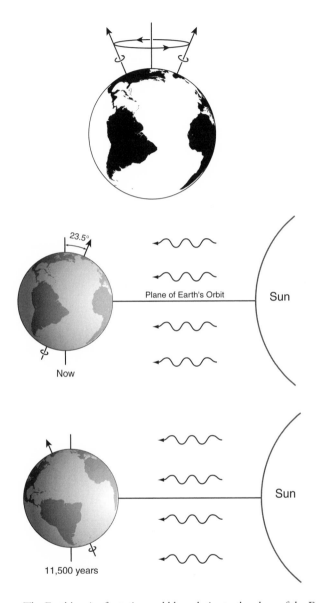

Figure 10. The Earth's axis of rotation wobbles relative to the plane of the Earth's orbit around the sun, just as a spinning top wobbles. The result is that for any particular point along the orbit, the direction of tilt gradually changes from year to year. One complete cycle takes approximately 23,000 years, so that half a cycle from now (in 11,500 years) the tilt will be opposite that of today.

cycle, the points along the Earth's orbit around the sun where the equinoxes occur—the fall and spring days when daylight and darkness have equal lengths—gradually change. But because the shape of the elliptical orbit also changes over time, the precession of the equinoxes follows a slightly different timetable than the wobble itself. One full cycle of this phenomenon is about 23,000 years, which means that every 23,000 years, the equinoxes occur at exactly the same point in the Earth's orbit. It's easier to think about this—and of more importance for glaciation—in terms of the winter and summer solstices. Today, the Northern Hemisphere has its longest day on June 21. That's the point along the Earth's orbit when the North Pole is tilted most directly toward the sun. But because of our wobbling axis, halfway through the precession cycle—11,500 years from now—when the Earth is at exactly the same point in its orbit, the tilt will be in the opposite direction (see figure 10). If there is anyone around in the Northern Hemisphere then, it will be December 21, the shortest day of the year and the beginning of winter in 13500 A.D. After another half cycle, 23,000 years from now, the tilt will be back to today's orientation.

But to return to Adhémar. He proposed that the wobble of the Earth's axis of rotation, combined with the eccentricity of its orbit around the sun, would cause the Northern and Southern Hemispheres to be alternately glaciated. His reasoning was that when the North Pole pointed away from the sun at the same time as the Earth was at its greatest orbital distance from the sun, the Northern Hemisphere would accumulate less heat and become covered with ice. Halfway through the cycle of the Earth's wobble, the same conditions would occur for the Southern Hemisphere. Adhémar did his computations assuming no change in the present-day elliptical shape of the orbit; the effects he predicted were entirely due to the wobble. He buttressed his arguments by pointing out that his theory predicted that there should be no current ice age for the Northern Hemisphere, because when the Earth is farthest away from the sun in its orbit, the North Pole points toward the sun, and it is Northern Hemisphere summer. Adhémar suggested that glaciation would occur again in the Northern Hemisphere only when the reverse is true, when the Earth is most distant from the sun during winter. As it turned out, there were errors in his calculations of the amounts of heat that would be accumulated in each hemisphere. But what really doomed Adhémar's theory were the wildly imaginative consequences he predicted. He claimed that at the cold pole, there would be a buildup of ice so massive that the Earth's center of gravity would shift toward it, catastrophically attracting the waters of the ocean in a kind of huge tidal wave. Every half cycle of the Earth's precession—every 11,500 years—as the ice built up at the opposite pole, the center of gravity would change again, with similar results. Adhémar envisioned a gigantic mushroomlike structure forming during the transition as the warming ocean water ate away at the base of the ice at the glaciated pole, leaving a huge cap supported on a thin column, which would eventually collapse, triggering the shift in the Earth's center of gravity and tidal waves crammed with icebergs. The land everywhere would be devastated.

Croll was influenced by Adhémar's book, particularly by the idea that the combined effects of eccentricity and the wobble of the Earth's axis might be important for glaciation. However, he showed that there were errors in Adhémar's calculations, which, in his view, invalidated some of the conclusions. Furthermore, "the somewhat extravagant notions which Adhémar has advanced," as he put it—namely, the collapsing ice pedestal and shifting center of gravity—were just a bit too fantastical for the serious Scot.

If others had already written about possible astronomical causes for the ice age, what then was Croll's contribution to this scientific problem, and why was it so important? The key, I think, was his determination to understand the process from first principles. His very first paper on the subject, which appeared in the August 1864 issue of *Philosophical Magazine and Journal of Science,* was masterful. He enumerated and evaluated each of the various theories that had been proposed to explain the ice age, and then summed up as follows: "Another objection which

we have to all these hypotheses which have come under our consideration is, that every one of them is irreconcileable *[sic]* with the idea of a
regular succession of colder and warmer cycles. The recurrence of
colder and warmer periods evidently points to some great, fixed, and
continuously operating cosmical law."

The geological evidence for alternating glacial and interglacial
episodes was by this time quite clear, and Croll realized that it could not
be ignored. "The true cosmical cause must be sought for in the relations
of our Earth to the sun," he went on to suggest. The fundamental principle, he recognized, is that solar energy is the external force that regulates the Earth's climate. For more than a decade, "the relation of our
Earth to the sun" was to be the focus of Croll's research.

Croll found that slight variations in the Earth's orbit around the sun
could not themselves be responsible for initiating or ending an ice age,
because the resulting fluctuations in solar energy received at the Earth's
surface were just too small to cause significant changes in average temperature. But unlike others who had reached a similar conclusion, he
did not immediately dismiss the idea. The cyclical nature of both the
orbital variations and ice ages convinced him that there must be a connection, so he turned to the problem of how the effects of small orbital
changes might be amplified. He began by examining how heat is distributed on the Earth, and concluded that ocean currents are extremely
important in determining the heat balance between the Northern and
Southern Hemispheres—important enough that changes in ocean currents could initiate or end glaciation in the Northern Hemisphere. The
orbital variations, he argued, while too small by themselves to cause
large temperature changes, could affect ocean currents and lead to glacial cycles. Secondly, having identified a mechanism, he carried out the
painstaking calculations referred to earlier, computing how the shape of
the Earth's orbit has changed over a period extending from three million years ago to a million years into the future. His graphs showed regularly alternating peaks and valleys, with many of the large peaks
recurring at roughly 100,000-year intervals (figure 11). The heights of

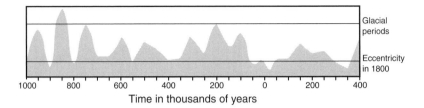

Figure 11. Part of James Croll's graph of the changes in eccentricity of the Earth's orbit around the sun. Croll believed that ice ages could only occur when the eccentricity was high, with values approaching the upper solid line on his graph. On this basis, he concluded that the most recent ice age had ended about 80,000 years ago. Graph based on data from Croll, *Climate and Time in Their Geological Relations: A Theory of Secular Changes of the Earth's Climate* (London: Daldy, Isbister, 1875).

the peaks varied considerably. Croll identified those times in his record when the orbit was most elliptical as the most likely glacial periods.

In what was really an extension of Adhémar's ideas, Croll reasoned that it was the interaction between the Earth's wobble and the changing eccentricity of its orbit that was most important. He asked himself what would happen if both the eccentricity was at a maximum *and,* because of the precession of the equinoxes, the Northern Hemisphere winter occurred when the Earth was farthest from the sun in its orbit. He concluded that the winters would be very much colder than at present. He also concluded that the cold temperatures would extend farther south, and that snow would fall and remain at more southerly latitudes than occurs today. This, in turn, would enhance the cooling, because snow reflects more of the sun's energy than bare ground. He also proposed that the greater temperature contrast between the equatorial and polar regions would increase the winds, which would affect ocean currents in such a way that the amount of heat carried northward by the Gulf Stream would diminish. The increase in wind intensity would also, he reasoned, carry more moisture from tropical regions toward northern latitudes, where it would precipitate and add to the accumulating blanket of snow. All of these factors—heat carried by ocean currents, the reflecting power of snow cover, and the availability of atmospheric moisture for precipitation—were first clearly demonstrated by Croll and are still recognized as important factors for the buildup of continental ice sheets.

Finally, Croll pointed out an obvious but often overlooked consequence of his theory: if he was correct, it could be used as a geological chronometer. When Croll did his work on ice ages, radioactivity had not yet been discovered, and geologists had no way to measure time accurately. The age of the Earth was unknown, although there were some wildly varying estimates based on such things as how long it would take an originally molten Earth's crust to cool to its current temperature. But the astronomical variations could be calculated quite accurately as a function of time. If changes in climate were really tied to the changes in the Earth's orbit, the geological effects—the glacial drift, the interglacial fossils and soils—could be dated simply by using Croll's graphs of orbital variations.

Although there are details of Croll's theory that were later found to be incorrect, it was nevertheless a remarkable achievement. This self-taught intellectual was the first to recognize the multiple interconnections in the Earth's climate system, something that is very much at the forefront of modern research. He anchored his theory in the astronomical variations, but realized that they are probably a trigger, rather than a cause, for glacial-interglacial cycles. In the most general sense, that is essentially the state of our understanding today.

Croll's papers on heat and electricity had brought him some recognition, but his astronomical theory for ice ages suddenly made him a very important figure in the world of nineteenth-century science. Fortuitous timing may have helped his rise to prominence, because there was intense interest in glaciation and its effects. The debate about the reality of an ice age had moved on; now the focus was on mapping the distribution of the glacial deposits, figuring out the sequence of glacial cycles, and, above all, determining the reason the climate had shifted in the first place. Croll was at the center of a hot field of enquiry, and the most eminent scientists of the day sat up and took notice.

Among those impressed by Croll's paper in *Philosophical Magazine* was Archibald Geikie, mentioned earlier, who had written extensively on glaciation in Scotland and who had just been appointed director of the recently reorganized Scottish Geological Survey. Geikie offered Croll a job. But Croll wrote to Geikie that he "did not see [his] way clear to accept the proposal." He was, he said, "somewhat up in years" (he was all of forty-three in 1864), but perhaps more important, he was not in the best of health, and he had no background in geological matters. He wasn't sure he had the necessary qualifications for the post.

Geikie was persistent, however, and he eventually persuaded Croll to move to the Edinburgh headquarters of the survey, not as a senior scientist, but as a "resident surveyor and clerk." Croll first had to take the Civil Service examinations—parts of which he failed—before his appointment became official, but in September 1867, he took up his new position. It was to be the last place of employment in his wandering career; he retired from the Scottish Geological Survey thirteen years later. During his time with the survey, he was showered with honors: an LL.D. degree from the University of St. Andrews, election as a Fellow of the Royal Society of London, and election as an Honorary Member of the New York Academy of Sciences. He was also awarded a number of scientific prizes. His official work was not very onerous, and he continued his "private work" in the evenings, writing papers and eventually a landmark book that brought together his ideas on climate change: *Climate and Time in Their Geological Relations: A Theory of Secular Changes of the Earth's Climate,* published in 1875.

Croll's influence on thinking about glaciation was at its peak. Many leading geologists took up his theory, recognizing that it was the best way to explain the geologically rapid alternation of glacial and interglacial periods that was implied by the field evidence. But as time went on and more evidence about the ice age accumulated, doubts began to surface. There were two main problems raised by the critics. The first was that Croll's theory, like Adhémar's, predicted that the hemispheres would be glaciated alternately every 11,500 years, half the length of the

Earth's wobble cycle. Secondly, Croll proposed that glaciation would only occur when the eccentricity of the Earth's orbit is at its maximum. Under these conditions, ice would form in the hemisphere that was tilted away from the sun—the winter hemisphere—when the Earth was also farthest from the sun in its orbit. Because the eccentricity cycle is much longer than the wobble cycle, eccentricity could remain large enough for each hemisphere to experience a series of glacial and interglacial cycles. As the eccentricity slowly decreased and the orbit became more nearly circular, the freeze-thaw cycle of the ice age would end, and the entire planet would have a warmer and much more uniform climate. Croll's calculations showed that the most recent period of high eccentricity was almost exactly 100,000 years ago, and that since then the eccentricity has decreased fairly rapidly to its present value. On this basis, he suggested that the last great glaciers must have retreated about 80,000 years ago.

Testing predictions is a key to scientific progress. Croll's theory could be tested if it were possible to date the deposits left by the ice age glaciers. The ages of glacial drift in the Northern and Southern Hemispheres should differ by 11,500 years, and if Croll was right about eccentricity, the youngest glacial deposits should be about 80,000 years old. Although there were no very accurate ways to measure geological time in Croll's day, there were some clever ideas. One was to use fossils. Species evolve over time, so that the assemblage of fossils in a deposit is generally distinctive for a given time. If the fossil assemblages in sediments at different places on Earth are similar, it generally means that they are about the same age. Fossils are not very abundant in glacial deposits, but sometimes, by combining fossil evidence and the principle of superposition, it is possible to bracket, or at least compare, the ages of several generations of glacial drift. When this approach was used for deposits in both the Northern and Southern Hemispheres, little evidence was found for different ages. In conflict with Croll's theory, there seemed to be no firm case for alternating glaciation in the two hemispheres. But the observation that really caused support for

Croll's theory to waver came from studies of erosion at Niagara Falls. Below the Falls is a steep-walled gorge, which has been formed by the waterfall's erosion of the rocks over which it flows. By direct observation, American geologists had estimated that the Niagara gorge is being eroded back at the rapid rate of almost a meter per year. This provided a way to estimate the age of the gorge, and, because the Niagara River flows across a deposit of glacial drift below the gorge, it also allowed a minimum age to be calculated for the glacial episode associated with the drift. The time estimated in this way was 10,000 years, clearly at odds with Croll's estimate that the last glacial episode had ended 80,000 years ago. Even though there were large uncertainties in the Niagara Falls calculation, and even though the age was later revised upward—to as much as 30,000 years—the discrepancy sowed seeds of doubt about the validity of Croll's astronomical theory of ice ages.

Croll still had many supporters, but they had to begin to make ad hoc arguments—perhaps, for example, local effects played a part, and thus the glacial records differed in Europe and North America. But it was a losing battle. In spite of the appeal of the astronomical cycles for explaining the alternation of warm and cold intervals, the predicted timing seemed to be wrong, and there was no conclusive evidence for alternation of hemispheric glaciation. Gradually, Croll's theory fell out of favor. By the end of the nineteenth century, the astronomical theory was all but dead.

At the beginning of the twenty-first century, however, Croll's astronomical theory is again alive and well. It has been refined and extended, but the basic premise is the same. Strangely, however, Croll himself is rarely mentioned, even in many modern textbooks. The orbital cycles, evidence for which can be seen very clearly in the changing sediments of deep sea cores, are usually referred to as "Milankovitch Cycles," named for the Serbian mathematician Milutin Milankovitch, who, to be fair, put the theory on a modern footing and is responsible for its renewed widespread acceptance. We shall see how he did that later in this book. Nonetheless, one has the feeling that Croll is a bit of a forgotten hero in the ice age story. Very occasionally, the astronomical theory of ice ages is referred to as the "Croll-Milankovitch" theory. Milankovitch, in his own writings, fully recognized Croll's contributions. Perhaps Croll will eventually get the recognition due to him from others as well.

CHAPTER SIX

Defrosting Earth

Joseph Adhémar's apocalyptic vision of a melting polar ice cap, a shifting center of gravity for the Earth, and devastating floods as ocean water rushed from one hemisphere to another is far from an accurate picture of what happens at the end of a glacial period. But at their maximum, the ice caps did hold huge volumes of water, and accumulating evidence shows that there *were* glacial floods as they melted, some of them catastrophic. In fact, some were so catastrophic that when the claim was first made that flowing water had produced the devastation they left in their wake, the idea was dismissed as preposterous. This reception was eerily reminiscent of the initial reaction to Agassiz's ice age theory. Catastrophic flooding associated with melting glaciers is now universally accepted, however, and in recent years, the topic has attracted renewed interest. Not only did the gigantic floods wreak havoc with the landscape, but they may also have been responsible for drastic shifts in climate.

The tale of gigantic glacier-related floods begins in the Pacific Northwest of the United States. In the state of Washington, there is a tract of land that early settlers called the scablands. It is harsh, dry land that to them was reminiscent of a partially healed wound, the soil ripped away and the countryside gouged and scarred by some unknown force. The unknown force, it turns out, was water from melting glaciers, water released in floods that were orders of magnitude larger than any known in human experience, so large that they completely transformed the landscape over which they flowed. Again reminiscent of the ice age theory, the idea that there was a connection between Pleistocene glaciation and the scablands was championed largely by one man, a University of Chicago geologist named J. Harlan Bretz. Bretz was a field man and an empiricist. Like many geologists, he considered his field evidence in the context of "multiple working hypotheses." The hypothesis that fit his data best, he concluded, was that the bizarre landscape of the scablands had been produced by one or more huge floods associated with the melting of ice age glaciers. Some of the features he studied could be explained by less dramatic processes, he admitted, but the only way to understand the totality of the scablands landscape was to invoke a flood of such magnitude that it was unique in the geologic record. In a paper written in 1928—five years after his first major published work on the scablands, and at a time when his ideas were being vehemently criticized by prominent geologists—Bretz wrote: "The region is unique: let the observer take the wings of the morning to the uttermost parts of the earth: he will nowhere find its likeness."

Bretz was born in Michigan in 1882. For a while he worked as a schoolteacher, first in Michigan and then in Seattle, Washington. It was there that he began his systematic studies of glacial geology, spending weekends and summer holidays tramping around the Puget Sound area and keeping detailed notes of his field observations. Eventually he decided to return to university to study for an advanced degree. Using the field notes he had accumulated in Seattle as the basis for a thesis, he graduated with a Ph.D. *summa cum laude* from the University of Chicago in 1913. Not long afterwards, he accepted a faculty position at his alma mater, where he was to spend the rest of his career. But living in the Midwest didn't divert his interest from the problems of glacial geology in the state of Washington. He quickly organized a summer field course for students, whom he took to the gorge of the Columbia River in southern Washington, where they studied, among other things, erratic boulders that they found distributed to elevations several hundred meters above the highest recorded water levels of the river. But unlike the erratic boulders of the Alps that were so important for Agassiz's theory, those along the Columbia River gorge could not have been carried by flowing ice—geologists had already established from moraines and other evidence that the Pleistocene glaciers had never reached that far south. The only plausible explanation was that the boulders had been carried to the gorge embedded in large floating rafts of ice that had broken off from glaciers hundreds of kilometers to the north. If that were the case, the distribution of the erratics would require very deep water, much, much deeper than the present day Columbia River. To explain the ice-borne erratics, some geologists concluded that this part of Washington had been submerged under the waters of the Pacific when they were deposited. But Bretz could find no evidence of ancient beaches or marine fossils, so he concluded that the icebergs had floated in fresh water. The first seed of the idea that there had been an unimaginably large flood as the Pleistocene glaciers melted away was planted in his brain.

Year after year for more than a decade Bretz took his University of Chicago students back to Washington for summer fieldwork. Especially in the early years, much of their travel was on foot. Later, Bretz had a car, which meant more ground could be covered. They were a motley crew: in addition to his students, Bretz was usually accompanied by his wife and children and the family dog. But motley crew or not, Bretz and his students gradually built up a detailed and intimate knowledge of the glacial geology of the area. After a few years concentrating on the Columbia Gorge in southern Washington, Bretz decided to move north to the scablands. Ever since his days as a schoolteacher in Seattle, he had been fascinated by the strange morphology of the landscape revealed on the topographic maps of the area—especially the giant "Potholes" waterfall, which sits several hundred meters above the present-day water level of the Columbia River. What Bretz and his students encountered in the scablands was, to Bretz at least, mind-boggling. To the untrained eye, it was a chaotic landscape of channels, basins, and hills. But when he had put all the pieces together, Bretz understood that he was looking at the work of a volume of flowing water so immense that it had eaten its way through hundreds of meters of loess, cut channels in solid basalt, and thrown up gravel bars so large that they made those of even the largest present-day rivers look positively Lilliputian.

Bretz added the adjective "channeled" to scablands, and ever since this vast area of eastern Washington has been known as the Channeled Scablands. He described it as a roughly rectangular block, bordered on the north and west by the deep canyons of the Columbia River, and on the south by the Snake. These two rivers merge in southern Washington and flow together westward to the Pacific. The Channeled Scablands are underlain by the dense, hard rock of the Columbia River Basalts, outpourings of lava that erupted in a flood of their own about sixteen million years ago and covered the preexisting landscape with layer upon layer of basalt flows. The Columbia River skirts along the northern and western edges of the basalt, but the floods that Bretz chronicled didn't follow the preexisting river course. Instead, they overflowed the banks of the Columbia and swept right across the basalt plain, scouring deep channels and potholes where none had been before.

Words and numbers can quantify the features that Bretz observed, but it is hard to grasp the magnitude of the floods that occurred without comparison to things that are more familiar. The floods swept across an area of the Columbia Plateau that is estimated to be at least 7,000 square kilometers. The peak water levels probably lasted no longer than days, possibly only hours. The amounts of water were prodigious—the best estimates suggest that more than ten *million* cubic meters of water per second coursed across the scablands. That is about sixty-five times the average discharge of the Amazon River today, and roughly ten times the average discharge of *all* the world's rivers to the oceans. And this deluge was both brief and confined to a limited area in eastern Washington. It's no wonder that the effects of the scablands floods baffled observers—no one had seen anything like them before.

Bretz must have arrived gradually at his conclusion that a huge glacial flood had created the Channeled Scablands. Certainly, the erratic boulders were an early clue, but it was the bizarre topography of the Scablands themselves that drove the message home. In his first published paper on the subject, which appeared in the *Bulletin of the Geological Society of America* and was innocuously titled "Glacial Drainage on the Columbia Plateau," Bretz described the scabland features, but he was cautious about interpreting them. He notes the fact that the Columbia River swings around the northern edge of the basalt plateau rather than cutting directly across it. He describes the geological history—the evidence that during glacial times, the Pleistocene ice sheets reached the northern border of the basalt plain but advanced no farther. He explains that during the ice age, the entire plateau was mantled with a covering of wind-blown loess about seventy meters thick, and that sometimes ash from volcanic eruptions in the Cascades Range to the west is mixed in with the loess. Then he focuses on the details of the scablands—the myriad intertwining channels that cut down through the loess to the basalt below, and, in places, extend far into the basalt itself; the deep dry canyons or coulees, some with walls that are three hundred meters high; the gigantic gravel bars that are made up nearly entirely of pebbles of the local Columbia River Basalt. He says that all of this must be the result of erosion by water, but he doesn't yet use superlatives, although he does refer to a glacial flood. He suggests that the water must have come from melting Pleistocene glaciers to the north, and that the ice must have dammed up the northern reaches of the Columbia River, because that would have forced the water to drain southward over the plateau rather than following the course of the river. Reading between the lines of his clear and measured prose, it is evident that Bretz already realized that the volumes of floodwater must have been enormous.

Over the several years following that 1923 paper, Bretz became increasingly confident in espousing and defending his theory that the

Channeled Scablands had been produced by catastrophic flooding. He continued to build up evidence from his field studies, and nearly every new observation seemed to bolster his theory. But there were many critics and very few proponents of his ideas. Geology by catastrophe was anathema to most geologists of the early twentieth century. The textbooks and conventional wisdom dictated that geological processes took place gradually, and that the carving of river valleys by streams was something that happened only over very long periods of time. In spite of Agassiz and his catastrophist ideas about the ice age, the gradualist view of nature had been in vogue for about a century, and for the most part, it had served geology well. Most geological processes *are* slow and gradual. Mountain ranges, the Grand Canyon, sand on a beach—these are all things that form over times that are very long from a human standpoint. Even the waxing and waning of the Pleistocene glaciers was understood by Bretz's time to be a gradual process, not the flash-freeze that Agassiz had envisioned, freezing woolly mammoths in their tracks. The Scabland floods, on the other hand, had to have been catastrophic. In Bretz's view, there was no alternative, and he saw no real contradiction in the possibility that the rates of geologic processes could occasionally change drastically. He realized that this would raise strong objections from those conditioned to thinking about erosion and deposition—the two processes that had created the unique landscape of the Scablands—as gradual processes. Once again reminiscent of Agassiz before him (although a bit more modestly), Bretz stated this explicitly when he wrote about his theory. In 1928, writing again in the *Bulletin of the Geological Society of America,* he observes that strong opinions can distort both sides of an argument: "Ideas without precedent are generally looked on with disfavor and men are shocked if their conceptions of an orderly world are challenged. A hypothesis earnestly defended begets emotional reaction which may cloud the protagonists' view, but if such hypotheses outrage prevailing modes of thought the view of antagonists may also become fogged."

Bretz rested his case on the field evidence. He would urge his critics—many of whom had no firsthand knowledge of the features he had described—to travel to Washington and view the Scablands themselves. Undoubtedly, he would have agreed with something Agassiz was reputed to have told his students: "Study nature, not books."

Bretz seems to have been a systematic and well-organized person. He was fond of lists. His papers often contain a succinct enumeration of all of the features he observed in the field. A partial inventory of the observations that led him to conclude that the Scablands had resulted from a catastrophic flood includes the following, paraphrased from his own lists:

· All of the areas of Scabland on the Columbia plain are connected to one another. The channels all slope downward from the northeast toward the Snake and Columbia Rivers to the south and west, and they are "anastomosing" or braided, unlike most river systems in which tributaries flow into a main stream in a dendritic pattern.

· There are just ten "openings" to the Scablands channels from the north, and nine exits where they drain into the Snake and Columbia Rivers.

· The loess, which still blankets other parts of the Columbia River Basalt plain, is almost completely stripped away in the Scablands tracts, and where it remains it is in the form of elongate, teardrop-shaped hills oriented parallel to the local Scabland channels.

· Although the basalt plain extends beyond the Columbia River in the west, and the Snake River to the south, no Scablands features exist there.

Even this partial list of Bretz's evidence for flooding is persuasive. Although on a different scale, the interfingering pattern of the channels in the Scablands—anastomosing as Bretz called it—is characteristic of most rivers that experience periodic flooding; they are aptly referred to by geologists as braided streams. The teardrop-shaped hills of loess in

the Scablands have steep sides that show signs of erosion by water, and
their overall pattern suggests a flotilla of ships, all with their prows
pointing upstream. The tops of these hills have soils developed on them,
and the surfaces of some of the larger ones have their own small but
normal drainage systems, completely unlike the braided channels that
separate them. Evidently, the floodwaters did not reach the tops of these
hills. The fact that all of the channels empty into the major rivers to the
south and west indicates that the floodwaters were eventually carried
out to the Pacific by the Snake and Columbia Rivers.

Bretz didn't use hyperbole, he simply presented his observations and
sought the best explanation for them. The scale of the Scabland features, he believed, required voluminous, short-duration flooding. For
the most part, no one questioned his observations, but his conclusions
were strongly opposed. In January 1927, he was invited to present his
views to the Geological Society of Washington, D.C. In the audience
were many of the most distinguished geologists of the day, and they had
choreographed a series of negative responses to what they anticipated
Bretz would say. Once again, he laid out his arguments. By this time, his
language had become more confident and forceful. Again he used bul-
letlike lists:

- Canyons of the Scablands. Largely channels of huge rivers.
- Rock basins in the channels. Thousands of them. . . . Lengths as
  great as eight miles, depths as great as 200 feet. . . . Formed by large
  vigorous streams plucking the . . . basalt.
- Cataracts. Hundreds of extinct waterfalls . . . several two to
  three miles wide.
- Trenched divides [here he is referring to elevated tracts of the
  landscape that separated different drainage systems but were
  "trenched"—cut across—by the floodwaters]. Several remarkable
  cases where . . . gashes 200 to 400 feet deep [cut] across a divide. . . .
  Water must have been 100 to 300 feet deep above the preglacial
  valley bottoms on the north to have crossed.

· 100 to 200 feet of loess removed over large areas.

· Contemporaneous occupation [by water] of all Scabland routes seems indicated. . . . Anastomosis due to the huge volume of glacial water and [its] abrupt introduction.

Although Bretz's arguments were impressive, there was one major weakness that many of his critics seized upon. Bretz was very much aware of the difficulty: he admitted that he had no reasonable explanation for the very sudden production of the huge amount of glacial meltwater required by his hypothesis. It was possible, he said, that a volcanic eruption had occurred under the ice to the north of the Scablands, resulting in massive melting, but there was no known independent evidence for such an event.

To Bretz, the source of the floodwater, although important, was secondary. He had walked over, mapped, and described in great detail the geological products of the flood, and he knew it had happened. But to his critics, more removed from the actual evidence, the source of the floodwaters was crucial. Coupled with their predilection for the slow, stately progress of geological processes, it was a potent reason to resist a catastrophic explanation. One by one they gave their rebuttals. The record of the meeting contains five lengthy statements from prominent geologists, all of them arguing that the features Bretz described did not require the type of flooding he postulated but could actually be the product of "normal" geological processes. One says: "Professor Bretz frankly points out the difficulties in applying his explanation of the origin of the remarkable features of the Columbia Plateau. It is not easy for one, like myself, who has never examined this plateau to supply offhand an alternative explanation of the phenomena . . . but I am left with the feeling that some things essential to the true explanation of the phenomena have not yet been found." Another critic comments about Bretz's estimate of the amount of water involved, concluding "criteria used to determine the actual quantities of water involved appear somewhat questionable." He goes on to say that Bretz's mechanism for producing this water by a subglacial volcanic eruption is "wholly inadequate." Yet another, discussing the deep channels in basalt that Bretz had described, comments: "Basalt is a hard rock and very resistant to corraision [a term for erosion by water that is carrying rock particles] . . . I am not convinced that so much work could be done on basalt in so short a time, even by such a flood as is postulated." Finally, another geologist who had actually seen Grand Coulee, one of the most spectacular of the great dry canyons of the Scablands, concludes: "The dry falls in the Grand Coulee resemble Niagara Falls and are evidently the product of normal stream work. The deep gorge of the coulee below the dry falls was apparently excavated by the same orderly and long-continued process . . . as the gorge below Niagara Falls, and it could hardly have been produced in a short time by a flood of whatever magnitude."

The published record of the meeting does not indicate whether anyone publicly supported Bretz's theory. Reading the comments of his opponents, however, it is evident that even in opposition, most of them were troubled by the sheer uniqueness of the Scabland features. Virtually all of them suggested that a lot more work in the region was required before a final pronouncement could be made about the processes that had molded the Scablands.

By the early 1930s, Bretz had effectively completed all of the fieldwork he would do on the Scablands problem. One very interesting and highly important piece of information came to light late in his studies, however. Examining the valleys and streams on the eastern and southern margins of the Scablands, Bretz found that all of them, including the largest, the Snake River, contained evidence for a surge of deep water flowing *upstream,* to the east, away from the Scablands. This reverse flood had left distinctive sediments along the valleys; in the case of the Snake River, Bretz traced these signs for almost one hundred and fifty kilometers upstream. The only reasonable explanation was that the water level in the Scablands had risen so fast and so high that it simply surged up all of the surrounding low-lying valleys. The force and magnitude of the flood had pushed water uphill, like a wave on a beach. It was further potent evidence for Bretz's catastrophic flooding theory. But still he was a lonely crusader. Throughout the 1930s, many field trips were organized to the Scablands, and other geologists made their own independent investigations. They floated alternative hypotheses: perhaps most of these features were the direct product of glacial ice flowing over the region (here they were ignoring the well-documented evidence that the ice sheets had terminated to the north of the Columbia Plateau); or perhaps ice had blocked up various key parts of the Snake-Columbia River drainage system, so that some of the region had been submerged under ponded water. If this were the case, rising water might occasionally have spilled over high points of land to create some of the Scabland features. Or perhaps nothing special was needed at all. There were still those who argued that the Scabland topography could be the result of normal stream erosion and sediment deposition. The debate continued quite fiercely during the decade, but Bretz never wavered in his conviction that a catastrophic flood had occurred—even though there was still no adequate explanation for the source of the water.

The turning point for the flood hypothesis came in 1940, in an unexpected way. That year, many of the principal players in the Scablands debate attended a meeting of the American Association for the Advancement of Science (AAAS) in Seattle, where there was a session on the glacial geology of the Pacific Northwest. One of the speakers, Joe Pardee, was a geologist with the U.S. Geological Survey who had worked in the Northwest for years. As early as 1910, Pardee had presented evidence of a huge, glacier-dammed lake in western Montana at the end of the last glacial episode—glacial Lake Missoula, named for the Montana city. The hills above Missoula are marked with multiple rings, easily visible today, that record former shorelines of the glacial lake. Some science historians have concluded that Pardee had realized decades before the Seattle meeting that Lake Missoula could have been the source of water for the Scablands flooding, and that he might even have told Bretz this. Surviving correspondence between the two men suggests that they may have discussed the idea, but if they did, it was never published. At the 1940 AAAS meeting, Pardee's presentation was about ripple marks in the sediments of the former glacial Lake Missoula. At first glance, that doesn't seem very catastrophic or exciting. Ripple marks are common enough features that most people have seen along a river's edge or in the sand of a beach. They are made by flowing water, and usually they are a few centimeters wide and about the same distance apart. If you walk over them in bare feet, it feels as though you are getting a foot massage.

In the title for his talk, Pardee put a question mark after the words "ripple marks." What he described were definitely not the small ripples on a beach; they were long, rolling ridges up to fifteen meters high and a hundred and fifty meters apart. Pardee commented that calling them ripple marks was really inappropriate, but he couldn't think of a better term. They occur in valleys near the western margin of the area that had been covered by glacial Lake Missoula, close to the border between Montana and Idaho and directly to the east of the Channeled Scablands. Pardee concluded that they must have been formed by the sudden release of a huge volume of water from the dammed glacial lake. He noted that it would have contained at least 1,700 cubic kilometers of water at its highest levels. Although Pardee didn't say so explicitly, it was quite apparent that a sudden release of Lake Missoula's water might have been the source of Bretz's catastrophic flooding.

Glacial Lake Missoula formed in the mountainous region of Montana now occupied by the Clarke River, which flows across the state from southeast to northwest. Glaciers of the Pleistocene Ice Age had blocked the normal drainage, and all of the valleys and low-lying areas over a vast region simply filled up with water, creating a lake with a complex, crenulated shoreline (figure 12). At its highest levels, the water crept up every little tributary valley in the mountains. It was about 650 meters deep at its deepest points.

The pressure of all that water was tremendous in the narrow valley where the ice blocked the natural drainage to the northwest. As the

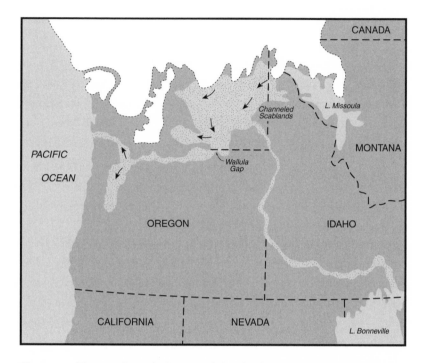

Figure 12.   The map shows the location of glacial Lake Missoula as the Pleistocene glaciers retreated into Canada. A lobe of ice blocked the northwestern drainage of the lake; when it gave way a large volume of water flooded westward onto the Columbia Plateau to create the Channeled Scablands. At the Wallula Gap, the water reached another bottleneck and ponded to great depths. Lake Bonneville, which also breached a (rocky) dam to flood the Snake and Columbia Rivers, is also shown.

climate warmed toward the end of the last glacial episode, melting ice added more and more water to the lake and the pressure on the dammed outlet increased relentlessly. One can imagine that one particularly hot summer, a trickle of water forced its way through the rotting ice dam. The trickle quickly became a stream, the stream a torrent. Suddenly, the ice dam gave way altogether, with a roar that must have shaken the countryside for hundreds of kilometers around. Lake Missoula poured out of its valleys and across the Columbia Plateau toward the Columbia River and the Pacific Ocean. Fish swimming in

Lake Missoula were floating belly up in the Pacific Ocean a few days later, if they hadn't been torn to shreds in transit. Soil, plants, trees, and boulders were caught up in the torrent and carried hundreds of kilomters. Elk, rabbits, and snakes were carried away in the roaring flood, which swept along everything in its path. Pardee calculated that the maximum outflow would have been about 31.6 *cubic kilometers* of water per hour. He noted that the Mississippi River, at the peak of its February 1937 flood, had discharged 0.16 cubic kilometers every hour. The Lake Missoula flow was nearly two hundred times as great. More recent calculations suggest that Pardee's estimate was too low, and that the actual peak flow may have been twice as high.

Pardee's observations removed the major objection to Bretz's flood hypothesis: the absence of a source that could supply a very large volume of water in a short time. The giant ripple marks implied just such a source, and the new evidence began gradually to win converts to Bretz's view. Still, some of his most vocal opponents took a very long time to come around. One, Richard Foster Flint, a widely acknowledged expert in glacial geology who had meticulously developed an alternative noncatastrophic theory for the Scablands landscape, found it particularly difficult to discard his own hypothesis. He had written *the* comprehensive textbook on glacial geology, but it was not until 1971, in his book's third edition, that he acknowledged the flood origin of the Channeled Scablands. Even then, he could not bring himself to be effusive about the unique character of the region. In a book of more than eight hundred pages, he devoted one dry paragraph to a discussion of the Grand Coulee and of Scabland features "widely created east of the Grand Coulee by overflow of an ice-margin lake upstream." He makes cursory reference to two of Bretz's publications.

Bretz lived to a ripe old age (figure 13). He returned to the Scablands one last time a few months before his seventieth birthday. By that time, there was abundant new information about the area, gathered through surveys by the U.S. Bureau of Reclamation, which had initiated a major irrigation project there. Bretz had access to Bureau of Reclamation excavations, aerial photographs, and new maps. One of the features revealed by the excavations was a complex pattern of layering in many of the Scabland sediment deposits, which suggested that there might have been many floods, not just one. That makes sense. The end of a glacial period is not abrupt; a few decades of cold weather and the glaciers would have advanced enough to again dam the exit from Lake Missoula, only to crumble as the warming continued, releasing yet another flood. Exactly how many occurred is still uncertain—some who have looked carefully at the evidence believe there may have been dozens, of varying size, over the several thousand years of the glaciers' demise.

At one place along the Columbia River's course, after it has been joined by the Snake River in southern Washington, there is a particularly narrow section called the Wallula Gap. It is far enough downstream in the Columbia River drainage system for all of the water from each of the Lake Missoula floods to have had to flow through this gorge—it is the only gateway to the west and the Pacific. Bretz found evidence that water had "ponded" behind this gap, backed up like a traffic jam at a narrow bridge. Initially, his critics were incredulous, because even at its highest levels, the present-day Columbia flows easily through the narrow valley. However, Bretz found signs that the floodwaters had risen to heights of at least 300 meters above today's valley floor and had backed up to flood the valleys of tributary streams such as the Yakima River for tens of kilometers. Wherever it ponded, the water, no longer traveling swiftly enough to hold its heavy load of clay and sand in suspension, began to deposit sediment. Each Lake Missoula flood added its contribution; today, in some low-lying regions of southern Washington, there is layer after layer of such sediment. In an otherwise harsh landscape, fertile, productive soil has developed on these patches of mineral-rich sediment, leaving an unexpected legacy of catastrophic ice age flooding: the burgeoning vineyards of southern Washington.

In 1952, in addition to finding evidence for multiple flooding, Bretz gained some startling new insights into the workings of the floods as he

Figure 13. J Harlan Bretz at age 95, at his home near
Chicago. Two years later, in recognition of his work on the
Channeled Scablands, he was awarded a prestigious medal by
the Geological Society of America. "All my enemies are
dead," he said, "I have no one to gloat over." Photograph
courtesy of the Special Collections Research Center, The
University of Chicago Library.

studied the Bureau of Reclamation aerial photos. With the bird's eye
perspective they offered, he found clusters of giant ripple marks in
many parts of the Scablands, just like the ones Pardee had described
near the exit from Lake Missoula. Although he had crisscrossed most of
these areas on foot during his early fieldwork, Bretz had never noticed
them. On the ground, pushing his way through sagebrush and a bit
like the proverbial flea on a camel's back, his close-up view had given

him no clue to the larger picture. The significance of the regular rise and fall of the land had not sunk in. The scale of the Scabland ripple marks, like Pardee's Lake Missoula examples, requires deep and fast-flowing currents. Bretz added yet another piece of evidence to his list supporting catastrophic flooding.

More recent research has shown that most of Bretz's early conclusions about the processes that formed the Scablands features were essentially correct. Residual doubts about the ability of water to very rapidly generate erosional features on the scale of those seen in the Scablands have been dispelled by engineering studies. Turbulent, high-volume flows have tremendous eroding power, especially when they are heavily loaded with rock particles. Vortices form in the roiling water that can act like a sandblaster, drilling into solid rock, and under some conditions, shock waves produced by breaking bubbles in turbulent flows can generate such large local pressures that they shatter almost any material. Even features as enormous as the Grand Coulee can be formed rapidly given enough water. With such knowledge, Bretz's work was entirely vindicated. Better late than never, the geological community recognized his contributions by awarding him the Geological Society of America's highest honor, the Penrose Medal, in 1979. He was 97. His one regret, he is reported to have confided to his son, was: "All my enemies are dead, so I have no one to gloat over."

There was a precedent for Bretz's flood hypothesis that was rarely mentioned during the debate. It involved another ice age lake in the western United States, albeit not a glacial lake. In 1890, G.K. Gilbert, a geologist with the U.S. Geological Survey, published a monograph on "Pleistocene Lake Bonneville," one of many large lakes that formed in low-lying regions of the west during times when the climate in that region was wetter than it is at present. Lake Bonneville occupied a large tract of land in northwestern Utah; the Great Salt Lake is but a small remnant of it. Gilbert's work describes features to the north of the former lake that appeared to him to be the result of flooding caused by a sudden release of water.

Lake Bonneville was too far south to have been dammed by ice, or to be supplied directly by meltwater from the major Pleistocene glaciers. But it filled and apparently emptied catastrophically toward the end of the most recent glacial episode, in the same time frame as the Lake Missoula floods. The northern exit from Lake Bonneville was blocked by loosely consolidated rocky rubble that filled in a low-lying area between hills. Just as the high water stand of Lake Missoula put enormous pressure on the ice dam that blocked its outflow, so a rising Lake Bonneville pushed mightily against its barrier of rubble. When it broke through, it rapidly cut away the loose material. To the north lay the Snake River plain. The waters of Lake Bonneville, reaching depths more than a hundred meters above the present valley floors, rushed northward into the Snake River and eventually, like the Scabland floods, emptied into the Columbia River and the Pacific Ocean.

There appears to have been only one major release of water from Lake Bonneville—once the barrier to its northern exit was eroded away, the lake could no longer fill up to such high levels. Estimates of the peak water discharge during the flood vary, but most place it at one-tenth of the maximum Lake Missoula rate, or less—still a very large flow. Potholes, channels, and flood deposits similar to those of the Channeled Scablands can be found for more than a thousand kilometers along the path of the flood. Gilbert's investigation focused on Lake Bonneville and its vast extent, rather than on the effects of the flood. His fieldwork was excellent and his writing clear, and there was little disagreement about his conclusions. Perhaps surprisingly, his descriptions of the features to the north of the ancient lake, and his interpretation that they were flood-related, did not attract the controversy that Bretz's work was to generate several decades later. Gilbert, like Bretz, realized that the landscape he was studying had been formed in a catastrophic process. That more traditional geologists did not immediately denounce this interpretation was probably because the scale of the erosional features, while significant, was much smaller than that of the Scablands, and because the source of the flooding was so obvious in the case of Lake Bonneville. G. K. Gilbert died in 1918. Had he been present for the debate about the Channeled Scablands, he would undoubtedly have been on Bretz's side.

A wonderful thing about science is its interconnectedness, and the fact that one discovery invariably leads to another, often quite unexpectedly. As the concept of catastrophic glacial flooding gradually gained acceptance, it became reasonable to ask other questions. Is there evidence for similar glacial floods elsewhere? Would the rapid addition of all that water to the oceans have had any important consequences? What about the sediment that was swept along by the floods? Could it be traced on the ocean floor? In the past few decades, all of these questions have been answered in the affirmative.

More than a thousand kilometers to the southeast of the point where the Columbia River empties into the Pacific, there is a ridge on the seafloor that has been studied intensively as a possible site of valuable mineral deposits. The Ocean Drilling Project, a multination project to study the ocean basins, has sent several expeditions to the area, and they have drilled long cores into the sediments near the ridge. In some of these, there are thick layers of sand that seem to have been laid down almost instantaneously. This is unusual, because normally, at this distance from land, the dominant type of sediment is a fine-grained mud. A little geological detective work on the cores showed that the mineral makeup of the sand closely matches that of modern sand from the Columbia River. It appears that the Scabland floodwaters, heavily laden with sediments, did not immediately mix with the seawater when they debouched into the Pacific, but instead continued to flow along the seafloor as dense "turbidity currents," only depositing their load of sediment when they eventually slowed down and spread out, very far from their source. The sandy layer closest to the top of the cores is sixty meters thick, and this is more than a thousand kilometers from the mouth of the Columbia! How far it extends over the seafloor is not known, but even if it is fairly limited, it represents an enormous amount of sedimentary material. This uppermost sandy layer is interpreted as being from the last major Lake Missoula flood, the same flood that gave the Scablands the form they have today. It seems that as each new aspect of the glacial Lake Missoula floods is uncovered, it underlines the gigantic scale and truly catastrophic nature of the processes involved.

The western part of North America was not the only site of flooding as the glaciers retreated. Signs of superfloods, as they have come to be known, have been found in northern Sweden, in Siberia, and in central Canada. Still, on the basis of available evidence, the Lake Missoula flood seems to have been one of the largest, although recent data suggest that it might be edged out of first place by floods that occurred in Siberia. As in the Channeled Scablands, the evidence for the Siberian floods includes deeply scoured channels, huge sand and gravel bars, and giant ripple marks. Like the Lake Missoula floods, those in Siberia were caused when an ice dam broke, releasing a huge volume of backed-up melt water from a glacial lake. Unlike the Scablands, however, the Siberian locality is mountainous, in the Altay Mountains near the northwestern border of Mongolia. Most of the water coursed along already existing river valleys rather than spilling over a flat plain and forming a new drainage system of its own, as happened on the Columbia Plateau. In the Siberian floods, the main erosive effect was that the river valleys were deeply scoured. The best estimates suggest that water levels in the main exit gorge for these floods reached 400 to 500 meters deep. Those who have studied the field evidence for the Siberian floods bill them as quite possibly "Earth's greatest floods."

However, the ice-dammed lakes that caused both the Scablands and Siberian floods were midgets compared to Lake Agassiz, the vast lake that formed along the margin of retreating Pleistocene ice sheets in central Canada. Its shorelines shifted around considerably during its more than 4,000-year-long history, but at one time or another, it stretched eastward from Saskatchewan across Ontario and into Quebec, and from Hudson Bay south into Minnesota and North Dakota. At its maximum, it is estimated to have held more than twice the volume of the Caspian Sea, the Earth's largest present-day lake. As the glaciers gradually retreated northward at the end of the last glacial period, drainage from Lake Agassiz periodically changed. Every so often, a new, lower outlet to the sea would become available and many thousands of cubic kilometers of water would be released quite suddenly. Over its history, the lake drained eastward through the St. Lawrence River to the Atlantic, southward through the Mississippi to the Gulf of Mexico, northwest along the Mackenzie River to the Arctic Ocean, and, finally, north through Hudson Bay. Lake Agassiz contained thousands of years of stored precipitation, and during its last major draining—nearly 8,500 years ago—it released an amount of water that was roughly one hundred times the peak volume of Lake Missoula. How catastrophic this event was is not known. Unlike Lake Missoula, Lake Agassiz didn't drain through a single, ice-blocked valley, however, but leaked into Hudson Bay through the margin of the gigantic retreating ice sheet. This may have softened its impact on the landscape (there is no visible surface evidence of flooding as there is in the Scablands or Siberia). Nevertheless, those who have mapped out the shifting shorelines of the glacial lake conclude that it drained suddenly, not in a series of small steps.

Even if the draining of Lake Agassiz didn't leave a record of giant potholes and ripple marks, it may have had another, quite different, effect: it may have influenced the ice age climate of Europe and North America, and perhaps the entire Earth. That is one of the reasons there is currently great interest in the history of this particular glacial lake. It is also another example of the fascinating interconnections that occur in science. How could Lake Agassiz, large though it was, have a profound influence on the climate of an entire hemisphere? The answer, in very general terms, is that the sudden introduction of a huge volume of fresh water into the ocean has the potential to change oceanic circulation—and oceanic circulation is closely tied to climate.

The Gulf Stream brings salty, relatively warm water into the North Atlantic Ocean, and as that water moves north and is cooled by the frigid arctic air, it becomes quite dense and sinks. That is what keeps the Gulf Stream operating: more water is continually drawn from the south to replace the water that sinks. But cooling the Gulf Stream is a two-way street; heat released from the water warms the air, and helps give Scandinavia, the British Isles, and indeed Europe as a whole, relatively mild climates. If the Gulf Stream circulation were to slow down, or stop altogether, the European climate, especially in the north, would resemble that of Siberia.

That may be what happened when Lake Agassiz suddenly drained into the North Atlantic. Introduction of such a large amount of low-density fresh water would have reduced the density of the North Atlantic surface water considerably, preventing it from sinking and thus slowing down (or perhaps even shutting off completely) the Gulf Stream. There is evidence that such a scenario may have occurred more than once. Each of the major changes in drainage that affected Lake Agassiz would have had a profound effect on ocean circulation. Ice cores from Greenland show that cold periods closely follow these drainage shifts, with two of the largest temperature drops in the past 15,000 years occurring near 12,800 and 8,200 years ago—approximately the times when Lake Agassiz released large amounts of fresh water into the North Atlantic, first when its drainage shifted from the Mississippi to the Great Lakes and the St. Lawrence, and later when it finally drained into Hudson Bay. Although it is difficult to trace past fluctuations in ocean circulation through direct measurements, these coincidences in timing constitute strong circumstantial evidence that there were such changes, and that climate was affected.

It may be that the interaction between melting glaciers and ocean circulation was more or less continuous, and that rapid release of meltwater from large lakes such as Lake Agassiz were simply short-term, speeded-up events in an ongoing process. As the climate warmed, ice melting would have supplied more fresh water to the North Atlantic, which would have slowed the transport of warm waters from the south, leading to a decrease in average temperatures. In the colder climate, retreat of the ice sheets would have halted temporarily, or the glaciers might even have advanced again. Eventually, as the overall global climate continued to warm, the ice would have begun to melt again. Such oscillations, driven by melting of glaciers around the northern rim of the Atlantic Ocean, would explain why there is evidence in the Channeled Scablands of multiple episodes of freezing and thawing of the ice dam that blocked the drainage of Lake Missoula.

The glacial floods chronicled by Harlan Bretz and others were cataclysmic events of the sort that occur only rarely, even on the very long timescale of our planet's history. It is quite clear from Bretz's early papers that he recognized the uniqueness of the Channeled Scablands and the processes that were responsible for their unusual morphology. But he most certainly could not have foreseen that half a century later, because of his work, he would become a guru for planetary geologists, and that his studies in the Scablands would be heavily cited in connection with similarly gigantic floods on another planet. Beginning in the 1970s and continuing up to the present, a series of spacecraft missions to Mars have beamed back images of features on the red planet that are remarkably like those of the Channeled Scablands. Although there are some dissenters, many planetary scientists are convinced that flowing water, some of it quite likely in the form of huge floods, have played a role in shaping the landscape of Mars. NASA has even sponsored conferences and field excursions to the Channeled Scablands in order to learn more about the processes that occurred during the Lake Missoula floods.

There is no running water on the surface of Mars today. At mid latitudes the average temperature is about $-75°C$, colder even than on the Earth's Antarctic Ice Sheet. In addition, the atmospheric pressure on Mars is only 1 percent of that on Earth, and there is almost no water vapor in the atmosphere. Any liquid water that appeared on the Martian surface would quickly freeze or evaporate into the atmosphere. Mars does have polar ice caps, however, composed of both water and carbon dioxide ice. There is also good evidence for water stored in the ground as permafrost, and—based on some of the most recent high resolution images—there may even be some real glaciers, partly covered with rocky debris.

In the simplest of descriptions, the geography of Mars comprises a southern highland region that has been heavily cratered by impacts, and a smoother, lower-lying northern section. The higher crater density in the southern highlands implies that this region is older than the less-cratered lowlands to the north. But some of the images that have been sent back from Mars suggest that the northern plains are smooth, not because they are less cratered, but because they are filled with sediment—there is evidence of craters and other features buried beneath the surface. Mars has frequent large dust storms—the very first mission to Mars, in 1971, encountered one that didn't clear to reveal the surface for months—so it is possible that the northern sediments were transported by wind. But it is also the case that some of the largest channels on the planet, with numerous features that suggest flowing water, seem to empty into the northern lowlands. Could this region once have been a sea of standing water, and could the sediment that now blankets the lowlands have been transported into this sea by giant floods? This is a question that has excited the imagination of both scientists and the general public, not least because the presence of liquid water on the Martian surface would enhance the possibility that extraterrestrial life has developed there.

In 1997, a spacecraft called the Mars Global Surveyor was put into orbit around Mars. On board is a camera that can resolve features on the surface down to about 1.5 meters in size. There is also a laser-based altimeter that can measure the height of the surface to a resolution of about 10 meters. Together, these instruments are giving planetary scientists a wealth of new information about the Martian landscape on a scale that was never before possible. The most recent data both confirm and extend earlier results that point to large-scale flooding at some time in Martian history. Information from the laser altimeter has been especially important, because this instrument makes it possible to map the elevation of Martian channels in great detail. Six of the largest channels on Mars—all of them estimated to be several billion years old—enter the northern lowlands and then suddenly become very indistinct. The laser altimeter measurements indicate that this occurs at about the same elevation for each of the channels, in spite of the fact that they are widely separated geographically. It has been suggested that this happens because the channels reached an ancient shoreline. They continue, indistinctly, far out into the northern plain, but these faint traces could have been made by the same type of sediment-laden "turbidity current" that carried sediments from the Missoula floods 1,000 kilometers out into the Pacific. If this interpretation is correct, if the channels really did empty into a standing body of water, it would have been a vast ocean covering 25 million square kilometers of the northern part of Mars to a depth of 560 meters. That would require a Martian climate very different from today's, even if, as some have suggested, the ocean persisted over a long time period only because its surface was frozen, keeping evaporation to a minimum.

The high-resolution images show that many of the Martian channels exhibit all the features of those in the Channeled Scablands. They often have an anastomosing pattern, widening and narrowing along their path. They flow around teardrop-shaped hills reminiscent of the residual loess hills in the Scablands. They show linear grooves and scouring marks similar to those produced in the Columbia River basalt by the Missoula floods. There is evidence of large cataracts. These features, especially when considered together, seem to require large-volume flooding. Estimates made by different scientists of the amount of water involved, however, range very widely, from channel flow rates not much more than that of the Mississippi River to truly catastrophic floods with peak flow rates nearly two hundred times greater than the largest Lake Missoula or Siberian floods.

One of the reasons why Martian flood volumes are so uncertain is that the source of the water is unknown. Just as Bretz's opponents seized upon this issue during the Scablands debate, so too critics of Martian flooding have focused on the source problem. There is no sign that there were glacier-dammed lakes above the Martian channels. However, some of them begin suddenly in regions of "chaotic terrain" that have been interpreted as areas where the ground was disrupted by sudden melting of subsurface ice, perhaps by volcanic heating. In fact, volcanic heating is the most plausible process for quickly melting large volumes of ice on Mars and initiating floods. Like Bretz, those who argue for catastrophic flooding on Mars point to the field evidence as proof enough, regardless of the source for the water, although for Mars, their "fieldwork" requires interpreting images rather than actually tramping over the ground and making direct observations. In spite of the similarities between the Scabland landscape and the features of the Martian channels, it will most likely require a breakthrough in understanding the source of surface water on Mars to convince the skeptics that catastrophic floods really took place there, just as it took Pardee's giant ripple marks to demonstrate that Lake Missoula supplied the water for the Scablands floods.

CHAPTER SEVEN

The Ice Age Cycles

Just a few years before Bretz published his first paper on the Scablands, a book appeared in Europe that was to have a lasting impact on thinking about the Pleistocene Ice Age. It was written by a Serbian mathematician who, in all probability, had never seen a moraine or an erratic boulder. But he was a gifted theorist who was ambitious and eager to take on large, unsolved problems in the sciences. The mathematician was Milutin Milankovitch, and he eventually settled on climate as a field that seemed ripe for the application of mathematical principles. Today, you will find his name in textbooks on geology, climatology, and the environmental sciences. Milankovitch revived and refined James Croll's ideas about the connection between ice ages and changes in the Earth's orbit around the sun. The repeated cold-warm cycles that are reflected in the advance and retreat of glaciers during the ice ages are now usually referred to as Milankovitch cycles.

Milankovitch and his twin sister were the eldest of seven children, born in 1879 into a well-to-do family in the town of Dalj, which sits alongside the Danube in what is now Croatia. The family had lived in Dalj for generations, and when Milutin was growing up, it was a pleasant and peaceful village, surrounded by fertile farmland, part of the Austro-Hungarian Empire. Milankovitch's education began at home, where he was taught by a governess, but at age ten, he was sent away to a secondary school in another town, where he stayed with relatives. "Sent away" is perhaps too strong a term; the school was only fifteen kilometers away, a shorter distance than people today travel to a favorite restaurant or a movie, but in Milankovitch's childhood, transport was slow, and he boarded with relatives. He was an indifferent student who found schoolwork easy and not very challenging, and he didn't work especially hard. Still, at the end of his second term, when the reports were being given out, he was called to the front of the class and proclaimed the top student. This came as something of a surprise to Milankovitch, and it placed upon him, he said later, a "moral obligation" to excel. And excel he did, especially in mathematics, graduating with top honors in the summer of 1896, at age seventeen. But then he was faced with a problem: What to do next?

The school that Milankovitch had attended specialized in science rather than following a more conventional classics-oriented curriculum. The choice of school was a conscious family decision; as the eldest son, Milutin would be heir to the family landholdings, and to manage these successfully, he would need a background in agricultural and technical sciences. But Milankovitch was not very interested in this future, so he made a pact with his brother: he would go on to study engineering, and his brother would study agriculture and then take over the family business. With the family's agreement, Milutin went off to Vienna to become an engineer.

Vienna, at that time the intellectual and cultural capital of the vast Hapsburg Empire, captivated the young student from the provinces. He took advantage of everything the city had to offer, especially its music, and he developed a lifelong passion for opera. He worked hard at his studies, too, but all too soon, his course in civil engineering was over, and he was sent to a military school in Zagreb to complete his mandatory military service. He was not happy with this turn of events; it was an experience that he thought might "kill all my human intelligence and independence and make out of me a robot." But instead of dulling his initiative, the change of circumstances gave him time to think about the future. Unhappy to be away from Vienna, he concluded that he didn't want to be a run-of-the-mill engineer in a provincial city. However, to take on really major problems in engineering and science, he would need additional education, which would probably mean returning to Vienna to pursue a doctorate degree—an appealing prospect. He would need money—the perennial problem of students— but he could rationalize that a doctorate would likely secure him a well-compensated position that would more than repay the investment. He found support from a generous uncle, and soon after his military service ended, at age twenty-four, Milankovitch was back in Vienna to begin his doctoral studies, determined to make a mark for himself.

Milankovitch was single-minded about his work. Just as he did not want to end up as an "ordinary" engineer in a provincial city, he also did not want to take on an "everyday problem" for his thesis. Original science was what interested him. And he also decided not to follow the usual course of doctoral students, which was to ask the most distinguished professors to suggest a thesis project. They would, he commented, give him only "the crumbs from their tables." Instead, he decided to find his own research topic, one that was both interesting and significant.

The subject he eventually settled on doesn't, at first glance, seem very glamorous, nor does it have any conceivable connection with ice ages. Milankovitch decided to work on . . . concrete. Although cement and concrete had been around as building materials at least since the Romans, concrete—especially the reinforced variety—was just beginning to see significant use in large structures. Whether or not Milankovitch realized that it would become the most important building material ever, he did feel that this was an area where he could both learn something new and also apply the principles of higher mathematics. Concrete was a material with known properties. He could calculate stresses and pressures and determine the size and shape limits that would be safe for actual structures. Not only would the mathematics be interesting, but the results would also have practical applications. Milankovitch dove into the problem with enthusiasm. It was his first real experience of independent scientific work, and he took to it like a duck to water. At the end of his project, when he had completed his thesis, he wrote in his memoirs: "I would be the happiest man in the world if I could live my whole life as I have this last year."

Milankovitch had been right about the value of his investment in education. The expertise in the properties of a new building material that he acquired during his doctoral studies gained him employment with a large engineering firm in Vienna. He was immediately given tasks that involved the design of large concrete structures throughout Europe. In less than half a year, he was appointed chief engineer, and soon he was patenting his ideas and publishing theoretical papers that dealt with problems in concrete construction. He had a comfortable life in a vibrant city, and a successful career as an engineer seemed to stretch out in front of him far into the future.

However, political events of the early twentieth century intervened. The "Bosnian crisis"—an earlier Bosnian crisis than the one of recent memory—cast long shadows across Austria. The Hapsburg Empire, after secret negotiations with Russia, annexed Bosnia and Herzegovina in 1908. This political maneuvering incited violent reactions from Serbia, which in turn provoked a backlash against people of Slavic origin in Vienna. Although he was not directly affected, the general atmosphere made Milankovitch uncomfortable. In the midst of this crisis, he was offered a position at the University of Belgrade, and, in spite of entreaties from friends and colleagues in Vienna—and the prospect of a drastic drop in salary—he decided to accept. Not only would he be going home, but he would also be able to devote himself to research in an academic setting. In 1909, he arrived in Belgrade to take up the chair of applied mathematics and begin a new career. It was quite different from anything he had done previously—his first task was to set up a complete three-year program for undergraduate students that would cover the various aspects of applied mathematics. Characteristically, he wanted to do things his own way rather than simply repeat the lessons given by his predecessor. He immediately and systematically set about drafting hundreds of pages of lecture notes. Although he had no experience as a teacher, Milankovitch was a good and popular lecturer. "I explained my mathematical formulae slowly and wrote them on the blackboard precisely," he said. He also interspersed his discussions of formulae with stories about famous scientists and the discoveries they had made, "which the students appreciated because they were not required to be learnt for the exams." Some things never change.

In spite of the fact that he had to live much more modestly in Belgrade than he had in Vienna, life as a university professor suited Milankovitch. He gave up his Austro-Hungarian citizenship and became a Serbian citizen. Once he had prepared his lecture notes, he turned to the activity he liked best: solving scientific problems mathematically. This is what he had done as an engineer; now he was free of practical constraints and could focus his talents on most any sphere of science. And although in most respects he was a modest and low-key man, in this sphere he was very ambitious. He was determined to make lasting contributions, and he sought out problems that, as far as he knew, were unsolved. Early on, one experience brought him up short. He had eagerly followed the scientific discussions surrounding Einstein's then-new theory of relativity, and he submitted for publication a short paper dealing with an aspect of the theory that he could treat mathematically. A short while later, he got the paper back with a note from the editor explaining that three separate papers covering the same ground had already been published by American scientists. Milankovitch had been completely unaware of them. It made him realize that to make an impact in a field, one had to be at the center of it and know the literature intimately. The University of Belgrade, while a good institution with some excellent faculty members, was clearly not a center of cutting-edge science. Milankovitch resolved to find an interdisciplinary area where he could use his expertise in mathematics to address important problems, "an arable field which I could cultivate with my mathematical tools," as he put it. Meteorology, he recognized, was just such a field. It was interdisciplinary, and nearly all the work being done was empirical. Like Croll before him, Milankovitch was intent on discovering the underlying principles, principles that he could investigate mathematically. In short order, his initial foray into the meteorology literature led him to climate and the problem of ice ages. He aimed very high: his goal was no less than to derive a mathematical theory of the Earth's climate.

By 1912, when Milankovitch first began to work on this problem, James Croll's astronomical theory for the origin of ice ages had been dismissed by most scientists because—as we have already seen—the timing he had predicted for glaciation did not match the estimated ages of glacial deposits very well, especially for the most recent glaciation, about which the most was known. Milankovitch read Croll's work, which he later called "remarkable." But initially he approached the problem somewhat differently, not so much from the perspective of finding the cause of ice ages, but rather of understanding climate in general from first principles. He started with the question, Is it possible to calculate, theoretically, what the temperature should be at any place on the Earth's surface? It was well known that the sun supplied the heat energy responsible for warming the surface, but calculating the actual temperatures would only be possible if he knew the values of all the parameters involved: the amount of energy from the sun that arrives at the top of the atmosphere, how this energy is transmitted through the atmosphere and spread over the Earth's surface, how much is reflected back into space, and the details of the Earth's position relative to the sun. The astronomical parameters—the Earth's position relative to the sun—were the same ones that Croll had investigated: the tilt and wobble of the Earth's axis of rotation, and its distance from the sun, which is controlled by the eccentricity of its orbit. As far as Milankovitch could tell from an extensive search of the literature, the detailed calculations he envisioned had not been done before. Croll had plotted the variations in the eccentricity of the Earth's orbit back several million years, and had attempted to show how the Earth's temperature might have changed in response to these changes, combined with the wobble of the rotational axis, but his calculations had necessarily been somewhat qualitative. He did not have accurate information about the amount of energy received from the sun, nor did he consider how heat energy is transmitted through the atmosphere. Milankovitch had the advantage of all that had been learned about these matters in the half century since Croll had first written on the subject. During that time, the amount of solar energy arriving at the top of the Earth's atmosphere had been determined quite precisely, and more accurate and detailed calculations of the variations in the tilt and wobble of the Earth's axis, as well as the eccentricity of its orbit around the sun, had become available for the period covering the past 1 million years of Earth history. Most of the important parameters that Milankovitch needed for his work were known or could be estimated quite precisely. The stage was set for him to begin his investigation.

The first major paper Milankovitch published on this subject appeared in 1913. In it, he gave a general account of his calculated theoretical relationship between heat supplied by the sun's radiation and the temperatures at various points on the Earth, taking into account the effects of the Earth's tilt and its annual orbit around the sun. It was the first real attempt at a theoretical portrayal of global climate, but it was only a beginning. Milankovitch knew that he had much more to do if he were to produce an accurate mathematical description of the Earth's climate. But in the summer of 1914, his life—and the lives of many in Europe—was thrown into chaos. Archduke Franz Ferdinand, the heir to the Hapsburg throne, was assassinated in Sarajevo by a Bosnian nationalist, and within a month, war was declared. Milankovitch, as a Serbian citizen, soon found himself in a prisoner-of-war camp in Hungary. He had recently married—the wedding had been just weeks before the assassination—and his new wife tried desperately to have him freed, initially with little success. Finally, she spoke to one of his professors from his days in Vienna, a well-known scientist who had powerful connections in the Austrian government. Milankovitch in prison was regrettable, but when his friend learned that imprisonment prevented him from doing science, he was truly dismayed. That was intolerable. The government was soon persuaded to release Milankovitch, although he had to agree to remain in Budapest and report weekly to the authorities. It was a very civilized arrangement, given the violence of the war, and it allowed Milankovitch to continue his work. He was welcomed at the Hungarian Academy of Sciences, where he toiled away on a massive treatise on climate. The manuscript was completed in Budapest; it was not until 1919, well after the war was over, that he was permitted to return home to Belgrade.

With a semblance of normalcy restored in Belgrade, Milankovitch could set about publishing his manuscript. It was an expansion of his earlier short papers on climate, a detailed and comprehensive treatment of how solar radiation affects the Earth's surface temperature at different latitudes and in different seasons. Because his theory didn't take into account the transport of heat around the globe by the oceans and the atmosphere, his calculated values didn't match the measured temperatures exactly—they were too low at high latitudes and too high nearer the equator. But averaged over the surface, they were in remarkably good agreement with the true, observed average temperature. This convinced him that his approach was the correct one—and equally important, it persuaded others too.

Milankovitch had written his manuscript in German, but in the end it had to be published in French. It was not a language in which he was fluent—he had to secure the help of a colleague from the University of Belgrade for the translation—but eventually, in 1920, his *Théorie mathématique des phénomènes thermiques produits par la radiation solaire* (A Mathematical Theory of the Thermal Phenomena Produced by Solar Radiation) appeared. Most of Milankovitch's earlier publications on climate and solar radiation had been short papers written in Serbian, and they had not been widely distributed. To some extent, his book—a thorough treatment of the subject in a language that most scientists in

Europe and many in North America could read—corrected this, although it was clearly not a best-seller. Milankovitch himself characterized its reception as "polite but lukewarm." He was not perturbed, noting that many important discoveries had gone unappreciated for years. If his work was significant, he said philosophically, it "would find its way without any help, recommendation or praise."

The book did attract considerable interest among meteorologists, because it was a departure from the traditional empirical approach to temperature and weather. For those who were willing to consider it seriously, it provided a theoretical framework for understanding meteorological observations. But in spite of the fact that it contained calculations of surface temperatures back to 130,000 years before the present, the book did not at first have much of an impact among geologists. A mathematical theory of temperature variations seemed remote from their immediate concerns. In fact, more attention was paid to Milankovitch's results for temperatures on Mars than to those for the Earth (he had carried out calculations for Venus, Mars, and the moon, in addition to the Earth). For several decades, the popular press had been full of stories of "canals" and sentient life on Mars, but Milankovitch's calculations predicted temperatures on the red planet far too low for liquid water or any life to exist.

Eventually, however, geologists and others interested in the Earth's climate history, especially its ice ages, came to realize the importance of Milankovitch's work. That appreciation arose through another of those serendipitous circumstances that are common in science and history. One of the meteorologists who read Milankovitch's book and was much impressed by his theoretical approach was the prominent German scientist Wladimir Köppen, who immediately saw the relevance of the calculations for understanding climate changes. It so happened that Köppen's daughter had married the geophysicist and arctic explorer Alfred Wegener, best known today as the first person to propose a comprehensive theory of continental drift. When Köppen read Milankovitch's book, he and Wegener were in the process of preparing a book manuscript on the subject of past climates. Köppen noticed that there was a direct correspondence between the timing of glaciation in the Alps (as far as it was then known) and the Northern Hemisphere temperatures calculated by Milankovitch. He immediately wrote to Milankovitch asking if he would be willing to collaborate with him and Wegener. Thus began a long association, which lasted until Köppen's death in 1940 and ensured that Milankovitch's calculations would be taken seriously. It was a fortunate and productive collaboration, because Milankovitch had the mathematical tools, Köppen had the international stature that guaranteed widespread attention to their ideas, and between them Wegener and Köppen had a good understanding of the geological evidence for ice ages.

Milankovitch's theory attempted to show how temperatures on the Earth would fluctuate over time in response to astronomical parameters. Like Croll before him, he considered the eccentricity of the Earth's orbit around the sun, and the wobble of its rotational axis, but he also introduced a new factor—he showed that the *amount* of tilt of the rotation axis is a far more important factor than had previously been believed. The angle of tilt doesn't change very much—it varies through a total range of only about two degrees, from roughly one degree less to one degree more than today's 23.5 degrees—but it markedly affects the contrast between winter and summer temperatures. This is not difficult to understand if you think about it. When the tilt is greater, the sun rises later and sets earlier in the winter hemisphere, and the converse is true for the summer hemisphere. So compared to the present, the summer hemisphere is heated more, the winter hemisphere less. Like the other astronomical parameters, changes in the amount of tilt are regular— they go through a complete cycle every 41,000 years. Milankovitch's work also revealed another aspect of the solar radiation effects that had not been realized previously. Because he made calculations for a range of latitudes rather than just for the Northern and Southern Hemispheres, as had been done before, he discovered that the various astronomical factors have quite different effects at different latitudes.

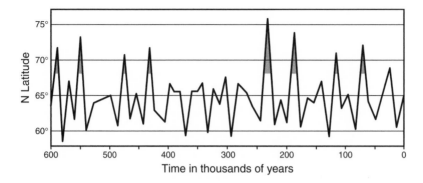

Figure 14. The graph that Milutin Milankovitch sent to Wladimir Köppen shows the "equivalent latitude" for 65° N in summer, based on the amount of solar energy received, for the past 600,000 years. Milankovitch believed that glacial intervals would occur at times when the equivalent latitude was significantly higher than today, indicated by the shaded peaks on this graph.

Köppen was especially interested in these nuances in Milankovitch's calculations. He argued that low summer temperatures in northern regions were the most critical factor for initiating permanent ice cover in Europe and North America—if summers were cool, more of the winter snow would persist without melting, the additional snow cover would cause more sunlight to be reflected back into space, and temperatures would drop further as a result, eventually leading to widespread glaciation. This was a radical departure from the conventional wisdom, the exact opposite of what most workers up to this time—including Croll and many of his supporters—had assumed: that cold winters were the key to the start of glaciation. According to Milankovitch's calculations, the same orbital conditions that produced cool summers resulted in winters that were slightly warmer than normal. But this did not bother Köppen—the higher winter temperatures would promote evaporation, and the temperatures would still be low enough that the resulting increased precipitation would fall as snow on the growing ice sheets.

In his book, Milankovitch had calculated temperatures at various latitudes on Earth through the past 130,000 years. However, all the evidence then available indicated that the Pleistocene Ice Age with its multiple glacial advances and retreats stretched back much farther in time than this. Köppen asked if Milankovitch could extend the calculations to at least 600,000 years before the present so that they could make a detailed comparison between theory and the field evidence. It would not be necessary, he said, to do the calculations for every latitude. He suggested restricting attention to the swath between 55° and 65° north latitude, since that was the region from which much of the detailed evidence for glaciation had been gathered.

Today, scientific results are frequently reported with the aid of elaborate, colorful graphs. Even in the 1920s, especially for interdisciplinary work, it was important to think about how the results would be presented. Milankovitch's calculations would result in tables of numbers that showed the amount of heat received from the sun at various latitudes. Tables could convey the information, but they would not have the immediate impact of a visual representation. Köppen and Milankovitch discussed the problem at length, finally deciding that a graph showing the changes over time would be best. That would illustrate for the reader both the magnitude of the variations, and exactly when the coldest and warmest intervals had occurred. When he had finished his calculations, Milankovitch sent Köppen graphs showing the amount of solar energy received at 55, 60, and 65 degrees north latitude, plotted against time (figure 14). He had calculated these values at intervals of 10,000 years back to 650,000 years ago, and it had taken him just over three months of intense work. Instead of plotting the actual solar energy values, or the estimated temperatures, which he knew were problematic because of the redistribution of heat by the ocean and atmosphere, he plotted what he referred to as the "equivalent latitude." This was an interesting idea because it showed immediately whether the latitudes he had chosen for his calculations received more (lower equivalent latitude) or less (higher equivalent latitude) solar radiation—usually referred to as "insolation"—than the same latitude does today. For example, Milankovitch's work indicated that

230,000 years ago, during the northern summer, the Earth received only as much solar energy at 65° N latitude as it receives today ten degrees further north. The equivalent latitude was 75° N (figure 14), indicating that the summer temperatures were much cooler at that time.

Köppen was pleased with the graphs. One of the reasons he had asked Milankovitch to calculate temperatures back to 600,000 years was that in the late 1800s, two German scientists, Albrecht Penck and Eduard Brückner, had developed a timescale for glaciation in the Alps that covered exactly that period. Based on observations of gravel terraces along Alpine river valleys, they had identified four glacial periods during this interval. Each terrace, they believed, corresponded to a period when active glacial erosion of the valley walls provided an abundant supply of gravel; when glaciation ended, the rivers could no longer transport much gravel and would begin to cut down through previously deposited sediments. As this was before the advent of radioactive dating techniques, they had to determine the age of each of the glacial periods by making assumptions about the rate at which the geological processes had operated. They simply measured the difference in elevation between successive terraces, and divided that distance by the rate at which they estimated the streams had eroded their beds. This was a very crude way to obtain the time that had elapsed between the formation of each of the terraces—and by implication, the successive glacial periods—but although most geologists recognized that the estimates were uncertain at best, real numbers published in papers by respectable scientists are seductive. Soon the four glacial periods described by Penck and Brückner became part of the vocabulary of European geologists, and the fact that American geologists had also mapped out drift deposits that seemed to come from four separate glacial advances gave their scheme added respectability. Even after other workers began to point out flaws in the Penck and Brückner field evidence, their version of the temperature history of Europe continued to be popular. It remained so into the 1940s before finally being discarded.

The Penck and Brückner scenario was very much in vogue when Köppen first saw Milankovitch's extended graph of equivalent latitudes. He realized at first glance that there was a close correspondence between the two—the times when Milankovitch's graph indicated cold Northern Hemisphere summers corresponded well to the times of Alpine glaciation based on the gravel terraces. The coincidence between these independent records—one theoretical and the other derived from field evidence—was very convincing; it seemed almost too good to be true. Milankovitch published the graph in a paper under his own name, and Köppen and Wegener put it in their book on past climates, with due credit to Milankovitch. The Köppen and Wegener book, *Die Klimate der geologischen Vorzeit* (Climates of the Geological Past), published in 1924, reached a wide audience and quickly brought Milankovitch worldwide prominence. The similarity between his graphs and the Penck and Brückner glacial timescale was so persuasive that he was generally acknowledged to have proven the astronomical theory of climate.

Milankovitch's graph soon became a cornerstone of much ongoing ice age research. Rather than testing his results further against other estimates of glacial timing, many geologists instead simply accepted them and began to use the graphs as a way to date glacial deposits. This was precisely what James Croll, many years earlier, had predicted could be done with his own graphs of the Earth's orbital variations.

Milankovitch did not rest on his laurels when his work became more widely known. Instead, over the next decade he continued to refine his calculations. He pushed them back even farther into the past, to a million years before the present. He expanded the range so that his results extended from 5° to 75° latitude, both north and south of the equator. As more precise data on the masses of the planets and their orbits became available, he incorporated them into his calculations of the Earth's orbital variations. Although these refinements did not materially change the conclusions that he, Köppen, and Wegener had drawn about past climate, they did provide a much more comprehensive picture of how the astronomical factors influence the solar energy budget over the Earth's surface through time. In the midst of this work, Wegener, who by this time had become a close friend, died tragically during a scientific expedition on the Greenland ice cap. The news of his death reached Milankovitch in May 1931, and although he had been expecting the worst—Wegener had set out on a dogsled journey the previous September and had not been heard from in the intervening months—it was a great blow. Wegener had been only fifty, and there was much he had still wanted to accomplish. His body has never been found.

Although Milankovitch's calculations and the conclusion that he, Wegener, and Köppen had come to about the connection between glaciation and insolation changes gained widespread acceptance, there was one question that continued to nag him. Beyond the general idea that glaciation coincided with periods of cool northern summers, he had no concrete criterion for the initiation of a glacial period. Were the insolation variations themselves sufficient? Or was there some other factor involved?

Milankovitch recognized, as had earlier workers, that the "snow line," the elevation at which permanent snow exists in the mountains, is an important indicator of average temperature. The elevation of the snow line varies with latitude—at present, permanent snow and ice occur at sea level at the poles, while in the tropics, they exist only on the highest mountains. Milankovitch realized that if the astronomical variations affected average temperatures—as his calculations showed—then they would also be correlated with the rise and fall of the snow line at a particular latitude. The difficulty was to find a mathematical relationship connecting the elevation of the snow line with his results for insolation. He realized, as had Croll, that there is positive feedback involved, because snow is much more reflective than forest or open ground, and as snow cover increases, more solar energy is reflected back into space. Fortunately, the appropriate measurements on snow and ice reflectivity had been made in the early 1930s, and Milankovitch could plug the relevant parameters into his equations with some confidence. The results were encouraging: the general pattern of his earlier curves remained, but the intensity of the cold periods increased when the reflective effects of the snow were included. When he calculated the altitude of the permanent snow line, he found that it descended to low elevations during the cold periods, and rose when it became warmer. In effect, a snow line near sea level meant that a permanent ice cap had formed. It appeared that the insolation variations, amplified by the effect of increasing snow cover, could indeed trigger the onset of glaciation.

With the inclusion of snow reflectivity in his calculations, Milankovitch had completed work on all aspects of the problem that he thought were important. But he had published the results of his investigations in bits and pieces, in different languages and at different times. His most famous graph was known by most people because of its inclusion in Köppen and Wegener's book. He was now corresponding with scientists from around the world, many of whom were working on the ice age problem, and most of them requested copies of his papers. There were no Xerox machines or electronic versions of his work, and he was quickly running out of the few copies his publishers had given him. He resolved to put all of his investigations of climate together in a book that would encompass everything, from the details of the equations necessary to calculate the motions of the planets around the sun to a discussion of ice ages. He called this massive work *Canon of Insolation and the Ice Age Problem*. It was written in German and published in 1941 (an English translation did not appear until 1969). Like Harlan Bretz, Milankovitch was very organized and systematic; his book begins with the classical laws of mechanics and then proceeds, step by step, through all of the calculations necessary to address the problem of climate variations.

Canon of Insolation was written during the early part of World War II, and it took Milankovitch, he says, 539 days (ever the mathematician, Milankovitch is precise; he doesn't say "about a year and a half," but gives us the time to the day!). The last pages were printed in Belgrade in April 1941; a few days later, Germany invaded Yugoslavia. Belgrade was bombed and the printing shop where his book was being produced was reduced to rubble. Fortunately, however, most of the printed pages

Figure 15.   Milutin Milankovitch working at his desk in 1954. Photograph courtesy of his son, Vasko Milankovitch.

survived intact. Even under occupation, some things continued to work in Belgrade—Milankovitch's rescued manuscript was bound and distributed, a brief happy occurrence in an otherwise unhappy time. His memoirs make it clear that his experience of life under German occupation was not pleasant. "Our civilized existence," he wrote, "had disintegrated into a life of hard grind."

Milankovitch was sixty-two and isolated from the world of science because of the war. The university was closed; his friend and colleague Köppen had died almost two years before, Wegener a decade earlier. His work on climate had been completed with the publication of *Cannon of Insolation,* and his scientific career was effectively over. Living conditions were grim, and the wait for the war's end seemed interminable. Not content to be idle, Milankovitch decided to write a history of science—an activity, he said later, that kept him sane. It was published in Serbian at the end of the war. After the Germans had left, life in Yugoslavia under the communists was only marginally better.

But Milankovitch had an international scientific reputation, and he was for the most part left in peace. He worked on his memoirs and some manuals for university courses (figure 15), and he was able to travel abroad occasionally. Late in 1958, at the age of seventy-nine, he died of complications from a stroke.

By then, the astronomical theory of glaciation was not in good health, either. This idea, which had seemed so promising when James Croll first proposed it, was, for the second time, losing influence among scientists who had initially embraced it. Milankovitch's work had revived the theory at a time when it had been all but forgotten, and his comprehensive calculations put it on a much firmer footing than Croll had been able to do. But once again issues of timing began to cast doubt on the theory. First, the gravel terraces of the Alps that had been thought to record glacial periods turned out to contain fossils incompatible with a cold climate. Later, it was discovered that the terraces had not been formed by glaciation, but had another origin altogether. The glacial timescale devised by Penck and Brückner, which had been one of the most important pieces of confirming evidence for Milankovitch's theory, was completely meaningless. As if that were not enough, there was a second blow to the theory when a new dating method showed that glacial deposits in North America did not correspond well in age to predictions of the theory.

The new technique was carbon-14 dating, invented in the late 1940s by Willard Libby and his students at the University of Chicago. It was an elegant method, making use of the fact that radioactive carbon-14 is continually produced in the Earth's atmosphere by cosmic rays, but then begins to decay away when it is incorporated into organic material. It had potential applications in a wide range of subjects, and Libby was later awarded the Nobel Prize in chemistry for his work. For the geologists working on ice ages, the new technique was a godsend. Here, finally, was a method that could provide "absolute" dates for the deposits of the glacial periods. Unlike other methods commonly used for measuring ages in geology, the carbon-14 technique cannot be used to date rocks; it works only for organic material that was once alive and exchanging carbon with the atmosphere. This meant that a fossil in glacial drift, or a piece of wood preserved in a peat bog were perfect samples for dating. There was one problem, however: radioactive carbon-14 has a half-life that, in geological terms, is quite short. Especially in the early days of its application, it could only be used to date materials that were not more than a few tens of thousands of years old. In older deposits, so much of the radioactive carbon-14 had decayed away that the few remaining atoms could not be detected.

Soon after Libby had demonstrated the feasibility of the carbon-14 method, others began to set up the necessary equipment, and before long, especially in the United States, there were a number of laboratories capable of making the analyses. Geologists studying ice ages had more than enough samples to keep them busy. Very quickly they established a timescale for the movement of the North American ice sheet by dating the moraines and drift that marked the margins of the ice at different times in its history. Even though they could not extend the analyses very far into the past, the carbon-14 data presented a far more complex story of glacial advances and retreats than Milankovitch's graph, which showed single cold-summer spikes at 25,000 and 72,000 years ago. History was repeating itself. Just as Croll's attempt to prove an astronomical cause for ice ages had foundered because the timing didn't seem to be right, so too the carbon-14 analyses in North America, especially in light of the complete abandonment of the Penck and Brückner timescale for European glaciation, seemed to sound the death knell for Milankovitch's revised and updated version of the theory. Furthermore, some meteorologists had looked again at the solar radiation balances calculated by Milankovitch and declared that the variations were just too small to produce drastic changes in climate, even when the effects of increased snow cover were incorporated. The pendulum was again swinging away from the astronomical theory. But the story has yet one more twist, one that brought Milankovitch's calculations back to center stage.

The evidence that revived the Croll-Milankovitch theory was discovered more than a decade after Milankovitch's death, and it came from the depths of the ocean. Many geologists had recognized that sediments on the sea floor might provide a long-term record of climate and other environmental conditions on the Earth's surface. Unlike moraines or loess or other glacial features on land, ocean sediments were presumed to accumulate slowly and continuously, century by century and millennium by millennium, without disturbance. If it were possible to core into these sediments, it might be possible to retrieve a record of events far into the past. Indeed, James Croll had presciently suggested that ocean sediments might provide the best clues about glacial cycles.

By the 1960s and 1970s, the technology for sampling the sea floor had improved to the point where it was possible to retrieve long cores of sediment reaching back millions of years into the past. Geochemists studying such cores noted that there are regular variations in the chemical composition of the sediments, and paleontologists examining fossils reported that there are similar alternations in the abundance of species that lived in warm and cold conditions. Oceanographers began to link these changes to glacial-interglacial cycles. But there was still the problem of assigning a timescale—accurate ages were needed to compare the dates of the sediments with the ages of glacial deposits on land and the timing of astronomical variations. Carbon-14 dating turned out to be more complicated for ocean sediments than for land deposits, and in addition there was the problem that it is limited to the past few tens of thousands of years. To probe farther into the past and investigate the cyclic changes that characterized the deep-sea sediment cores would require a new approach.

The breakthrough that came to the rescue occurred in another, unrelated, area of earth science. Geophysicists studying the Earth's magnetic field discovered that it has reversed periodically in the past—the north and south magnetic poles have switched positions. When rocks form on the Earth's surface—when lava erupts from a Hawaiian volcano, for example—the magnetic minerals they contain line up their own small magnetic fields in the same direction as the Earth's. By dating such rocks and measuring their magnetic orientation, a record has been built up of how the Earth's magnetic field has varied in the past. Particularly important is the timing of the magnetic reversals, the geologically short intervals when the field switches from "normal" (today's situation) to reversed, because these serve as markers or time lines that can then be used to date other events. Through the work of many different laboratories, the reversals have been dated quite accurately.

Ocean sediments, like the igneous rocks of Hawaii, contain magnetic minerals. As they slowly settle to the sea floor, these minerals too line up with the Earth's field. The long cores from the oceans thus contain a continuous record of changes in the magnetic field, a built-in timescale for dating variations in sediment properties that might be related to glacial climate cycles. Many of these properties seemed to vary in a regular way, but the question was, Which would be most useful for understanding ice age climate? The answer was not immediately obvious, but one property in particular turned out to be crucial for confirming, once and for all, the link between climate and variations in the Earth's orbit. It was the oxygen isotopic composition of fossil shells in the sediments. What in the world, you may wonder, do oxygen isotopes have to do with glaciation or the astronomical theory of climate? The connection was made by Harold Urey, a chemist who, like Willard Libby, worked at the University of Chicago, and who, again like Libby, was awarded the Nobel Prize, his for the discovery of deuterium, one of the isotopes of hydrogen. Urey was especially interested in how different isotopes of the same element behave when they take part in chemical reactions or processes such as evaporation and precipitation. Most of the elements in the periodic table have multiple isotopes; oxygen, for example, has three. All three have the same chemical properties—they are all oxygen—but they exhibit minute differences because of their different masses. In a tank of oxygen gas all of the molecules have the same energy. They whiz around, bumping into the walls of the tank and each other, but those containing oxygen-16 travel slightly faster than those containing oxygen-18, because oxygen-16 is lighter.

From theoretical considerations, Urey discovered that during chemical reactions, the oxygen isotopes are fractionated from one another because of the slight differences in their masses. One isotope is preferred over another in the products of the reaction. Furthermore, he found that the amount of fractionation depends on the temperature. In a flash of insight, he realized that oxygen isotopes could act as a natural thermometer. Many organisms that live in the oceans make their shells of calcium carbonate, an oxygen-containing compound. The oxygen comes from seawater, and when the shells are being precipitated, one oxygen isotope is preferred over another. Thus the shells end up with different proportions of the three oxygen isotopes compared to seawater, and the amount of that difference depends on the water temperature. By measuring the ratio of oxygen-16 to oxygen-18 in an ancient shell, Urey realized, he could determine the temperature of seawater in the distant past—a stunning concept.

Like most such ideas, bringing this possibility to fruition took some time. Urey and his students had to perfect the measurement techniques so that they could measure oxygen isotope proportions accurately in small amounts of calcium carbonate shell. On the basis of their calculations, the variations would probably only amount to a few tenths of 1 percent. They also faced the perennial problem of geochemists: Which samples would be most representative and provide the most important information? Their method, too, they soon learned, had its own complications. For example, they realized that both evaporation and precipitation would change the proportion of the various oxygen isotopes in seawater. How could they distinguish these variations from those caused by temperature changes?

By the early 1970s, most of these difficulties were well understood. Fortunately for research into the Pleistocene Ice Age, evaporation of water from the oceans—the process ultimately responsible for the supply of snow to the glaciers—and cold seawater temperatures both act to change the oxygen isotope proportions in carbonate shells in the same direction. Although it was still difficult to separate the two effects in a quantitative way, they at least reinforce one another, producing a stronger oxygen isotope signal than would temperature changes alone.

Research groups interested in glaciation and the Earth's climate history scrambled to make oxygen isotope measurements on deep-sea sediments. Two papers from these early studies were especially important in the debate over the astronomical theory of climate. The first was by Wally Broecker and J. van Donk, who were working at what was then the Lamont-Doherty Geological Observatory (now the Lamont-Doherty Earth Observatory) of Columbia University. Published in 1970 in the journal *Reviews of Geophysics and Space Physics,* their paper had the title "Insolation Changes, Ice Volumes, and the Oxygen-18 Record in Deep-Sea Cores." Broecker and van Donk used the magnetic properties of the sediments to determine a timescale for the cores, and they showed that when their oxygen isotope analyses were plotted against this timescale, they exhibited a smooth and systematic variation over the past 400,000 years.

What was puzzling about this graph for advocates of the Milankovitch astronomical theory was that it showed several cycles of peaks and valleys, with each cycle lasting about 100,000 years (figure 16). James Croll had predicted that the eccentricity of the Earth's orbit, with a cycle close to 100,000 years, would be important for glaciation, but Milankovitch's calculations of Northern Hemisphere temperatures had shown that the more important parameter is actually the tilt of the Earth's axis of rotation. The tilt changes through a cycle lasting approximately 41,000 years. Combined high eccentricity and maximum tilt might result in especially severe glaciation, but according to Milankovitch, the tilt should be the determining factor. Why didn't the oxygen isotopes follow the tilt cycle rather than exhibiting regular 100,000-year variations?

Broecker and van Donk's work was not the only study that showed the 100,000-year cycles. Several other groups, some approaching the problem from different perspectives, found the same thing. Was the eccentricity really the important thing after all, or was there some unknown process at work with a 100,000-year cycle not connected in any way with the Earth's orbit? The debate about whether or not astronomical variations could be responsible for glaciation heated up once again.

As more and more oxygen isotope data accumulated, it became clear that the same variations found by Broecker and van Donk were present in deep-sea cores from all of the world's oceans. They were a global phenomenon, and they had to reflect the temperature changes and waxing and waning of the ice sheets that characterized the Pleistocene Ice Age. The magnetic timescale was crucial for this conclusion, because sediments accumulate at different rates in different places. The only way to be sure that the peaks and valleys in the oxygen isotope records occurred simultaneously throughout the globe was through accurate dating of each core using its magnetic properties.

The 100,000-year oxygen isotope cycles discovered by Broecker and van Donk are quite regular, but they are not perfectly smooth. Superimposed on these long cycles are many smaller wiggles. In a few cases, early workers examined cores that extended back to a million years or more, and in these they found that the prominent 100,000-year cycles seemed to die out between about 800,000 and one million years ago; before that there were also cyclical variations, but on a shorter timescale. How could all of these features of the oxygen isotope graphs be explained? Astronomical changes still seemed attractive because of the regular nature and planetwide occurrence of the variations. But once again the question of reconciling the timing of glaciation with the well-determined orbital variations became an issue.

The problem was solved in 1976, in a paper that appeared in the journal *Science*. Using the technique of spectral analysis—an approach that is capable of disentangling multiple, superimposed, cyclical curves and retrieving the original characteristics of each type of cycle—James Hays, John Imbrie, and Nick Shackleton showed that the oxygen isotope record is actually made up of several distinct, superimposed cycles,

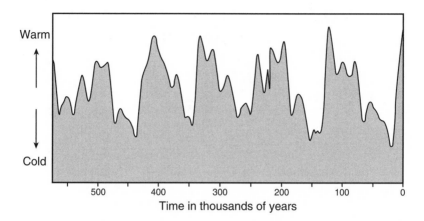

Figure 16.   A representation of climate changes over the past 550,000 years, based on oxygen isotope analyses of deep-sea sediments. The oxygen isotope "proxy" combines information about both temperature and the amount of glacial ice that exists on the continents. It is obvious that cold and warm periods alternated on a roughly 100,000-year timescale over this period.

with timescales corresponding almost exactly to the predictions of the astronomical theory: a 100,000-year cycle that reflects changes in the eccentricity of the Earth's orbit, another cycle of about 43,000 years, close to the timescale of changes in the tilt of the rotation axis, and one near 20,000 years that corresponds well with the wobble of the rotation axis. Hays, Imbrie, and Shackleton titled their paper "Variations in the Earth's Orbit: Pacemaker of the Ice Ages." Finally, it seemed, the ideas that had been formulated first by Croll and then refined and extended by Milankovitch had been shown to be correct. Exactly how the orbital variations get translated into glacial-interglacial temperature differences was still uncertain, because virtually every analysis that had been carried out concluded that these variations result in only very small changes in the amount of solar energy received on Earth. But the close correspondence in timing between the astronomical cycles and the isotopic properties of deep-sea cores could not be denied. It was unlikely to be a coincidence. The repeated buildup and decline of the vast

Pleistocene ice sheets must have been linked directly to changes in the Earth's position relative to the sun. The Earth's orbit is truly a pacemaker for climate.

CHAPTER EIGHT

Our Planet's Icy Past

Oxygen isotopes in deep-sea cores, together with a few other indicators of past climates, have given us a surprisingly clear picture of the coming and going of glaciers during the Pleistocene Ice Age. But the current ice age occupies only the past few million years, an almost insignificant slice of our planet's four-and-a-half-billion-year history. What was the climate like for the rest of that vast sweep of time? The "norm," if it is possible to speak of such a thing, was one of warmth and little or no permanent ice. However, there is good evidence that our small (by the standards of the universe) planet has experienced sporadic ice ages for at least the past three billion years. Almost as soon as Louis Agassiz had pointed out the significance of glacial drift and other ice-produced effects in the 1830s, geologists began to find similar signs of ice ages in the more distant past. The very first such reports came from India, where, trapped within layers of sedimentary rocks, deposits of glacial drift were found lying atop scratched and grooved bedrock. Unlike Pleistocene drift in the Alps or in Canada, the ancient drift in India was not loose, but had been cemented and indurated into solid rock over hundreds of millions of years. Such drift-turned-to-rock was termed "tillite" by geologists, employing a seventeenth-century word describing chaotic rock deposits that contain fragments of a variety of sizes.

Soon similar occurrences of tillites had been found in Australia, South Africa, and South America. To early geologists, one of the most startling aspects of these discoveries was that many of the places showing evidence of past ice ages were tropical or subtropical. In those pre-plate-tectonics days, when the continents were believed to be fixed and immobile, it was difficult to imagine tropical ice sheets. It seemed reasonable enough to think that glaciers in the Alps had once been more extensive, or that there might have been ice in Scotland in the past, but an Earth with glaciers near the equator was hard to grasp.

Aside from the problem of ice in the tropics, a major difficulty for those attempting to characterize ancient ice ages was that much of the evidence is missing. Even for the Pleistocene glaciation, erosion has obliterated some of the geological signs of ice action. The most recent glacial advance of the Pleistocene ended only some twenty thousand years ago, and most of its effects on the landscape are still quite obvious. But earlier Pleistocene glacial advances and retreats, even those that occurred only one or two hundred thousand years ago, are much more difficult to study, because the moraines and erratics and scratched bedrock from those episodes have not all been preserved intact. Such difficulties are compounded many times over for the ice ages that occurred in the Earth's very distant past. But in spite of this, geologists have been able to identify at least four periods of severe glaciation that occurred long before the Pleistocene, all probably more intense than the current ice age. The timing of these is shown schematically in figure 17—the earliest known dates to about 2.9 *billion* years before the present, the most recent, 300 million years. Some of the early ice ages are depicted here as a series of events stretching over several hundred million years; at these distant times in the past the uncertainty in dating is such that it is not clear whether these were actually discrete ice ages, or just especially severe intervals within an overall cold period. In addition to the ones shown, several other times have been identified when the planet experienced cool periods, if not full-blown ice ages.

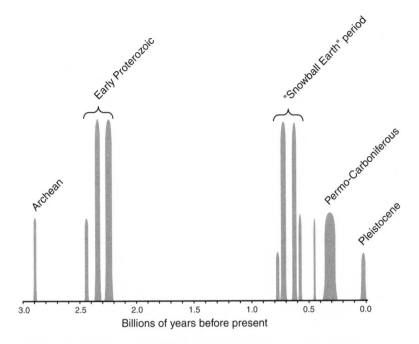

Figure 17. The Earth's major ice ages, as identified from glacial drift, tillite, varves, and glacially scoured bedrock. Heights of the shaded bars give a rough indication of the intensity of these glacial periods, although the estimates are speculative for the glaciations between 2.2 and 2.4 billion years ago. In addition to the four major ice age periods before the Pleistocene discussed in the text, a further cold period that occurred 450 million years ago has been identified and is shown here. It is not possible to represent the durations of ice ages accurately on this small graph.

Traveling back into Earth's history from the present, the first really major ice age that appears in the geologic record occurred about 300 million years ago. This is the very same ice age for which evidence was uncovered in India and other southern continents in the nineteenth century. It is worth considering for a moment what criteria must be satisfied before an ancient event can be called an ice age. How is it possible to know, for example, that a tillite, or glacial scratching, results from global glaciation and not from local mountain glaciers? Usually, at least three important characteristics must distinguish the evidence for glaciation.

First, the effects of ice sheets should be widespread, usually meaning that they occur on several, well-separated continents. Secondly, the widespread glacial deposits must be contemporaneous. And, finally, there should be evidence that the glaciation took place at low elevations in most localities. This is not so difficult to establish as it might seem, because when ice sheets reach the sea, they drop glacial drift into the ocean, where it is preserved under later blankets of sediments.

The older the purported ice age, the more difficult it becomes to satisfy all of these criteria, especially the criterion of contemporaneity. If a dating method is accurate to a few percent, the uncertainty in dating a glacial advance that occurred 100,000 years ago is only a few thousand years, but for a 300-million-year-old tillite, it can be six to ten million years. That's several times the length of the entire Pleistocene Ice Age. Furthermore, no method has yet been devised that can accurately date a glacially scratched surface, or an ancient tillite. Usually, the best that can be done is to bracket the age by dating lava flows or volcanic ash layers that occur above or below the glacial deposits. Sometimes, fossils in sedimentary rocks that accompany the glacial deposits are useful too, but usually these can only bracket the time of glaciation and do not date it directly.

The ice age that occurred near 300 million years ago is sometimes referred to as the Permo-Carboniferous Ice Age, after the two geological periods that it spans, the Permian and the Carboniferous. There is evidence that the Earth was cold for about 80 million years, from 340 until 260 million years ago. During this very long span of time, there were many cycles of glaciation and deglaciation, much as has occurred during the current ice age, although because of their great antiquity, it has not been possible to work out the timing of these cycles with any confidence. Whether or not they were influenced by astronomical cycles is also not known, but given the importance of orbital variations for the Pleistocene Ice Age, it is likely.

One of the interesting aspects of the Permo-Carboniferous Ice Age is the role it played in ideas about continental drift. When Alfred

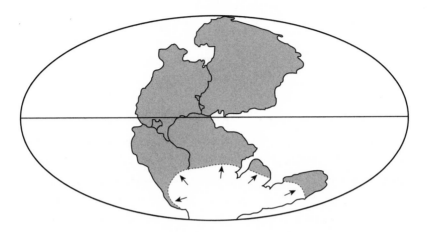

Figure 18. At the time of the Permo-Carboniferous glaciation, the Earth's land masses were joined in the supercontinent Pangea, which stretched from the South Pole to northern latitudes. The current southern continents—Africa, India, Australia, and South America—were clustered together with Antarctica near the South Pole, in a landmass referred to as Gondwanaland. Ice sheets spread northward to at least 40° south latitude.

Wegener—the same Wegener who worked together with Wladimir Köppen and Milutin Milankovitch on past climate and ice ages—was putting together his theory that the continents had moved about over the Earth's surface, he searched for geological features that appeared to be continuous across now-separate continents. He was especially struck by the widespread evidence for the Permo-Carboniferous glaciation. Glacial deposits had been reported from all of the southern continents—India, Southern Africa, South America, and Australia—and as far as could be determined, the glaciation had been roughly contemporaneous. Wegener realized that a single, very large, continental ice sheet could account for all of the deposits if these now widely spread localities had once been contiguous. That would require closing up the Atlantic Ocean so that Africa and South America were joined, pushing India up against the eastern coast of Africa, and somehow attaching Australia to this group of continents. In addition, the problem of explaining how glaciers could exist in the tropics would disappear if his theory of continental drift were correct—the great agglomeration of landmasses could have been located much farther south, near the South Pole, at the time of glaciation, and only later drifted to their present, much warmer, localities.

Although it turned out that Wegener was right about mobile continents, his theory was heavily criticized and did not immediately settle the controversy over the Permo-Carboniferous glaciation. Wegener was quite selective in his choice of evidence for continental drift, and he simply ignored things that seemed contradictory. He also could not identify a satisfactory mechanism for moving the continents about—his one suggestion, that centrifugal forces associated with the Earth's rotation might be responsible, was quickly demolished by his opponents. As a result of these difficulties, the theory was generally discounted, and with it the idea that the Permo-Carboniferous glaciation occurred when the southern continents were joined together and located near the South Pole. It was not until almost half a century later that the concept of continental drift was resurrected and transformed into the modern theory of plate tectonics. But present-day reconstructions of the southern continents at the time of the Permo-Carboniferous glaciation show that Wegener's conclusion was essentially correct. Africa and South America had been joined together like pieces of a jigsaw puzzle, with India snuggled up against Madagascar on the east coast of Africa. The Antarctic continent and Australia were firmly attached and nestled along the southern tips of Africa, South America, and India. This entire supercontinent, referred to by geologists as Gondwanaland, had been centered near the South Pole throughout the long Permo-Carboniferous Ice Age. When locations of the glacial deposits are plotted on the fused-together continent, they reveal the former presence of a single large ice sheet (figure 18).

The development of the theory of plate tectonics also solved a related puzzle. Glacial markings often reveal the direction of ice flow, and some of the striations and scratches left by the Permo-Carboniferous glaciation indicate that the ice flowed from the sea onto the land, at least in terms of the present-day configuration. In India, for example, they show ice moving inland from the Arabian Sea. The evidence is unambiguous, but because ice flows downhill under gravity, the conclusion appeared to early geologists to be impossible. Glaciers form on land and flow into the sea; how could great thicknesses of ice build up in the ocean and spread onto the land? But with the continents joined together as Gondwanaland, there were no intervening seas. Ice simply flowed outward from its thick center across the present-day continental boundaries, creating the impression—after the continents were separated—that it had flowed inland from the oceans.

The most reliable dating of the Permo-Carboniferous Ice Age suggests that its maximum extent occurred during a span of about 20 million years, from approximately 280 to 300 million years ago, and that ice sheets extended over the supercontinent of Gondwanaland to at least 40° south latitude, and perhaps even to 35° S or less. That equals or exceeds the maximum spread of the Pleistocene glaciers in the Northern Hemisphere during the current ice age, and it is reasonable to infer that the volume of ice at the height of the Permo-Carboniferous glaciation exceeded the maximum, so far, of the Pleistocene glaciation.

But how do we know this? How can the latitudes of, say, Buenos Aires or Cape Town three hundred million years ago be determined, when we know the continents have continuously moved about on the Earth's surface? The *relative* positions of the continents can be worked out quite far back into the past, but locating them with respect to the poles is a more difficult problem. And yet estimating the extent of ice sheets, particularly estimating how closely they approached the equator, is crucial for determining the severity of an ancient ice age. Fortunately, the Earth's magnetic field provides a tool for this problem, just as it does for the timescale of deep-sea sediments. The field is very close to one that would be produced if there were a gigantic bar magnet inside the Earth. Viewed from space, it would look just like the textbook example of iron filings sprinkled around a bar magnet: the field lines form great arcs that join the magnet at its top and bottom and bulge outward at its sides. And at every latitude, the field has a specific orientation relative to the Earth's surface, ranging from 90° at the poles to 0° at the equator.

As was discussed in the previous chapter, some minerals line up their internal magnetic fields with the Earth's when they form, just as the iron filings line up with the field of a bar magnet. And because the orientation of the Earth's field depends on latitude, such minerals essentially encode the latitude in their physical properties at the time of their formation or when they are incorporated into a sedimentary layer. In favorable cases, the magnetic information can be deciphered in the laboratory, and although it's not always possible to tell which hemisphere the rock formed in (the inclination of the magnetic field to the surface is the same at equivalent latitudes both north and south of the equator), it is possible to determine the critical parameter for glaciation, how close it was to the equator. Like most geological records, the evidence from rock magnetism gets more and more fragmentary as one delves farther and farther back into geological time, but for the Permo-Carboniferous glaciation, the data are abundant. Not only can the positions of the continents that made up Gondwanaland be located quite precisely, but it's also possible to reconstruct how the supercontinent drifted slowly across the South Pole during the long ice age. The record left by the ice itself is fully consistent with the magnetic evidence, for it shows that the center of the ice sheets—the region from which ice flowed outward in all directions—stayed roughly fixed in latitude near the pole as the continent drifted slowly through the region.

Today's northern continents seem to have largely escaped the Permo-Carboniferous glaciation. This too is consistent with the magnetic evidence, which indicates that they were not then located in polar regions. Only parts of what is now Siberia extended to high northern latitudes; North America and Europe were farther south. There are some signs of localized glaciers in Siberia, but in Europe and North America, the tillites, glacially scratched rocks, and other ice effects that exist in the former Gondwanaland are absent. But there is one striking feature of the geologic record in these regions that has been linked, indirectly, to glaciation: an abundance of coal deposits.

The Carboniferous period derives its name from the widespread carbon-rich deposits that occur in this interval of geologic time. They are mostly made up of the remains of spore-bearing plants similar to ferns, plants that lived in low-lying, moist environments often referred to as coal swamps. The organic debris that accumulated in these swamps formed peat deposits; these in turn were eventually transformed into coal. A peculiar feature of the Carboniferous coal deposits in North America and Western Europe is that they are cyclical: beds of coal alternate with marine sedimentary rocks such as limestone or shale in a pattern that is repeated many times over. In places as many as a hundred cycles occur; although it is difficult to determine the amount of time represented by each cycle, in aggregate it is estimated that they span ten million years of deposition, or even more. The plants that were the precursors of the coal grew in fresh or slightly brackish water, and it is believed that many of the coal deposits formed in low-lying coastal swamps that were periodically inundated with seawater. With each flooding the accumulated peat was buried beneath a layer of ocean sediment; between floodings the fresh water swamps reestablished themselves and new layers of peat accumulated. The pressure and heat of burial eventually transformed the multiple peat layers into the cyclical coal deposits that characterize the Carboniferous.

The link between coal deposits and the Permo-Carboniferous Ice Age has to do with the repeated flooding of the coal swamps, which was most likely due to rising and falling sea level. All glacial ice has its source in the ocean through the Earth's ongoing cycle of evaporation and precipitation; if all the water currently frozen in the Greenland and Antarctic ice caps were returned to the ocean, sea level would rise by about 60 meters, more than enough to drown a coastal swamp. At the height of the Permo-Carboniferous glaciation, the amount of water locked up in the ice sheets was probably equivalent to somewhere between 150 and 250 meters of ocean depth. Even if only a fraction of these ice sheets melted and then reaccumulated in glacial-interglacial cycles similar to those of the Pleistocene Ice Age, the resulting changes in sea level would be quite sufficient to explain the cyclical Carboniferous coal deposits. The cause of the alternating warm-cold periods of the Permo-Carboniferous Ice Age is unknown, but their regularity hints at an astronomical or other external "pacemaker," just as it did for the current ice age.

The Permo-Carboniferous Ice Age was long and severe, with vast regions of the southern continents buried under ice that extended from the South Pole to low latitudes. Although life on Earth was not nearly as diverse 300 million years ago as it is now, the existing fossil record shows that it was significantly affected. Throughout the Gondwanaland super-continent, the diversity of plant life dwindled, and the species that survived were hardy varieties adapted to harsh climatic conditions. But as severe as the Permo-Carboniferous glaciation was, it has gradually become apparent that there was an earlier period in the Earth's history that was even worse. Although there is lively debate about just how much worse it really was, there is general agreement that it was probably the most pervasively cold era in our planet's history. Some researchers contend that the entire planet was frozen—not only was there ice on the continents, but the ocean surface was frozen as well. This has been called the Snowball Earth hypothesis. Those who opt for a slightly less extreme climate refer to Slushball Earth. Regardless, it was intensely cold, much colder than the Earth has ever been in human experience.

Snowball Earth occurred about 300 million years before the Permo-Carboniferous glaciation, during the long interval between 550 and 850 million years ago. Several separate ice ages may have occurred during this period, but the uncertainties in dating glacial deposits and the dif-ficulty of correlating from continent to continent mean that the entire interval is usually referred to as a single ice age. It occurred near the end of the Proterozoic eon of the geological timescale, and to distinguish it from other icy episodes, geologists refer to it as the Late Proterozoic glaciation. The Late Proterozoic Earth was a very different world than the one we know today—plants and animals had not yet appeared on land, and only relatively primitive life inhabited the sea. The continents were mostly barren rock, the oceans contained no fish or lobsters or seaweed, and there is good evidence that even the atmosphere was quite different, with much less oxygen than at present. Because the Late Proterozoic Ice Age happened so long ago, the evidence of glaciation, although very strong, is fragmentary. It remains only in places where it could be easily preserved, usually places that were already submerged at the time of glaciation, or were low-lying and later flooded by the sea as the glaciers receded. In such environments, glacial drift, scratched and scoured bedrock, and other glacial features were buried under sediments and stored for hundreds of millions of years—and later uplifted again for geological inspection today. But in spite of the rather stringent conditions required for its preservation, the evidence for glaciation during the Late Proterozoic is widespread. It is found on every continent, suggesting that ice sheets were present throughout the globe.

The widespread distribution of glacial features has long convinced geologists that this ice age was unusually harsh. At least two and perhaps as many as four or five major icy episodes, separated by warmer intervals, have been identified during the long cold period. But the big surprise about the Late Proterozoic glaciation came in the late 1950s and early 1960s, when researchers found that the magnetic properties of rocks associated with some of the glacial deposits indicated formation at very low latitudes. As we have seen, during the present ice age, Northern Hemisphere glaciers have never pushed farther south than 40–45°, and even for the more severe Permo-Carboniferous Ice Age, there is no indication that ice reached much closer to the equator than 35° latitude. In both cases, fossil evidence suggests that the tropics remained fairly warm. If Late Proterozoic glaciers had existed near sea level in the tropics, that would indicate a very different ice age indeed, and initially many geologists were skeptical. But as more and more data became available, the initial results were corroborated. It seemed inescapable that frigid climates had extended very close to the equator. Thus was born the concept of Snowball Earth.

The idea that there was a truly global ice age, a Snowball Earth, is still controversial. Geological evidence suggests that during most of our planet's history, the Earth's average temperature has stayed within rather narrow limits, and critics ask, Under what conditions could this have changed so that the Earth froze over completely, from pole to equator? And, if it happened once, why haven't we had more Snowball Earth episodes? At the extreme of Snowball Earth conditions, the average surface temperature would have been closer to that of Mars than anything we are familiar with today—perhaps about $-50°C$. Could even the primitive life of the Late Proterozoic have survived such extremes? There is general agreement that the glaciation was severe, but as this is written, the jury is still out on whether or not a true Snowball Earth occurred. It is nonetheless worth examining some of the major issues in the controversy.

The magnetic data are perhaps most crucial for the hypothesis, because they fix the latitude of the continents at the time of glaciation. The term "Snowball Earth" itself was coined because magnetic measurements placed the glaciated regions in the tropics, so it is reasonable to assume that their reliability has been carefully scrutinized. It is well known that magnetic measurements can be problematic, especially for old rocks, because they rely on a very accurate determination of the "frozen in" magnetic orientation that was captured when the rocks formed. With care, the orientation can be measured quite accurately with respect to the rock's current position on the Earth's surface, but what happens if the rocks have been folded, or tilted, or otherwise moved from their original positions at some time over the past 700 million years? And what if they have been deeply buried and heated, as often happens? Heating can have a significant effect on the stored magnetism, and in extreme cases, heated rocks can be remagnetized— perhaps in an orientation that is completely different from the original. Fortunately, there are ways around these problems. Those who study the magnetic properties of rocks have devised ingenious laboratory approaches that allow them to strip away the magnetic "overprinting" caused by heating and to recover the original magnetic orientation. And because the layers of sedimentary rocks are always horizontal when they are first deposited, folded and tilted sedimentary rocks can be "unfolded" or "untilted" (not literally, but by applying a correction factor to the measured data) to recover their original orientation. Doing this for several samples tilted or folded at different angles actually provides a very good test of the reliability of the measurements: if they all agree after unfolding, one can have a high degree of confidence in the results.

Magnetic measurements have been made on rocks from around the world that are associated with the Late Proterozoic glaciation. They have been repeated in different laboratories with good agreement, and they show that the majority of these localities were at latitudes less than 30° at the time of glaciation, and none were farther than 60° from the equator. In fact, there seem to have been no landmasses near either of the poles in the Late Proterozoic.

Even if glaciers existed at low latitudes, there is still the question of whether the evidence we have comes from ice in high mountains, or at low elevations. This is not such a difficult question to answer, because, as already mentioned, most of the glacial effects that remain from the Late Proterozoic are preserved in sedimentary rocks. It is fairly clear that some of the tillites were either deposited in shallow seawater at the edge of a continent, or so close to the shoreline that a slight change in sea level engulfed them in marine sediments. The same is true of preserved glacial markings on bedrock. Overall, the evidence is very strong that even the tropical ice sheets extended right down to sea level.

However, even such extreme climate conditions do not automatically lead to the central conclusion of the Snowball Earth theory: that the oceans were frozen too. Evidence for that idea came first from an examination of ocean sediments from the Late Proterozoic by Joe Kirschvink, a geochemist at CalTech, who coined the term "Snowball Earth." In 1992, Kirschvink pointed out that peculiar sedimentary deposits rich in iron, referred to as banded iron formations, or BIFs, occur in a number of localities around the world just at the time of the Late Proterozoic glaciation. Geologists were familiar with BIF deposits from very early in the Earth's history, but none were known for about a billion years before the Late Proterozoic glaciation, and none have formed since that time. Their occurrence requires the buildup of very large amounts of dissolved iron in seawater, a phenomenon that cannot occur today because of the oxygen-rich atmosphere. As anyone who has had to deal with rusty metal knows only too well, oxygen combines rapidly with iron and forms rust. BIFs are basically rust deposits (with a few other components as well), and their restriction to the early part of the geologic record is thought to be due to the low concentration of oxygen in the atmosphere at that time. Under such conditions, iron from the weathering of both continental and undersea rocks would accumulate in the oceans until it came into contact with oxygen— perhaps produced by photosynthetic algae living in surface waters— whereupon it would be oxidized and precipitate out as a BIF. The occurrence of BIFs in the Late Proterozoic was an enigma, because by that time in the Earth's history, there was enough atmospheric oxygen to prevent the necessary buildup of dissolved iron in the ocean. But Kirschvink reasoned that if the ocean were frozen, preventing any exchange with the atmosphere, its oxygen content would be rapidly depleted. Iron concentrations would increase to high levels, and BIFs would be deposited when the sea ice melted and oxygen from the atmosphere again began to exchange with the ocean.

Kirschvink's proposal seemed reasonable, but it was not convincing to everyone. Perhaps locally oxygen-poor basins—which occur because of restricted circulation even in today's oceans—could have served as hosts for BIFs during Snowball Earth. That would still not explain why they are absent before and after the Late Proterozoic, but it did cast some doubt on the frozen ocean hypothesis. Then, in 1998, four Harvard University researchers, led by the geologist Paul Hoffman, published the results of a study they had made in northern Namibia, which, they believed, made a strong case for the Snowball Earth theory. The region showed clear evidence of low latitude (approximately 12°S) glaciation between about 760 and 700 million years ago. An interesting and important aspect of their work was that the rock sequence they investigated indicated that the glaciation had both started and ended abruptly. Immediately overlying the glacial deposits—as is the case at many other Late Proterozoic Ice Age localities—they found limestone-like sedimentary rocks of a kind that form only in warm, tropical waters. Where they occur over glacial deposits, these distinctive formations have been termed "cap carbonates" by geologists. They signify a rapid transition from very cold to very warm conditions.

The presence of cap carbonates in Namibia highlighted one of the issues consistently raised by critics of the Snowball Earth hypothesis, the problem of how the Earth could ever have thawed out again once it was completely ice-covered. The high reflectivity of the snow and ice would have bounced much of the solar energy that normally warms our planet right back out into space. An entirely frozen Earth, critics of the theory claimed, would have reached a climatic point of no return and could never have recovered. But the cap carbonates in Namibia and other localities indicate that it did recover, and very rapidly at that. Had the deep freeze really been as severe as the proponents of Snowball Earth would have it? And if it was, just how had the climate changed abruptly from icy to tropical?

A possible solution to the permanently frozen Earth problem had actually been suggested by Joe Kirschvink several years before the work of Hoffman and his colleagues. Kirschvink's idea was that carbon dioxide, a "greenhouse" gas that traps solar energy in the form of heat in the atmosphere, would build up to high levels under Snowball Earth conditions, eventually leading to global warming and complete melting of the glaciers. His reasoning went approximately as follows. First, we know that the main source of CO_2 to the atmosphere is volcanic eruptions, which spew out gases as well as lava. Secondly, there are two removal mechanisms that keep CO_2 roughly in balance—one is photosynthesis (plants use CO_2 to make organic tissue, and release oxygen into the atmosphere as a by-product), and the other is the chemical weathering of rocks on the surface. This process will be explored in more detail later in this book, but in the simplest terms, CO_2 is removed from the atmosphere because it dissolves in rainwater to make carbonic acid, which attacks and dissolves rocks. Because there is no evidence to suggest that the level of volcanic activity—the source of carbon dioxide—was radically different in the Late Proterozoic compared with today, a buildup of atmospheric CO_2 could only occur if there was a decrease in its rate of removal. That is exactly what one would expect if the oceans were frozen—most of the photosynthetic organisms living in the sea would die, greatly diminishing one of the removal mechanisms, and the cycle of evaporation from the ocean and precipitation over land that drives the other, chemical weathering of rocks, would also cease. Dry, desertlike conditions would prevail globally. With the two primary CO_2 removal processes greatly diminished or shut down altogether, its concentration in the atmosphere would be expected to rise to quite high levels.

The CO_2 concentration that would be required to warm the Earth and melt a completely frozen ocean today is very high. At the time of the Late Proterozoic Ice Age, even higher contents would have been necessary. According to astronomers who investigate the life cycles of stars, our sun's energy output has gradually increased over the Earth's history and would have been significantly lower during the time of Snowball Earth. Some researchers believe this was a factor in the initiation of the Late Proterozoic ice ages, but whether it was or not, a CO_2 concentration several hundred times that of today's atmosphere would have been required to thaw the completely frozen planet. At those levels, melting would begin first at low latitudes, and, once begun, would proceed in a runaway fashion with the help of positive feedback. Decreasing ice cover meant that more and more of the incident solar energy warmed the ocean instead of being reflected back into space. Evaporation from the newly uncovered ocean increased atmospheric humidity, which in turn intensified the greenhouse effect, because water vapor is an even more efficient trap for the sun's energy than CO_2. If this version of the Snowball Earth theory is correct, not only did the planet suffer through periods of extreme cold in the Late Proterozoic, it also endured brief but intense "super greenhouse" episodes when global temperatures soared far above anything experienced since, before returning to more normal levels as the CO_2 balance was restored.

The work in Namibia by Hoffman and his colleagues supports the Kirschvink scenario of a buildup in atmospheric CO_2. A centerpiece of their investigation is the chemical data they collected. Like those who study the effects of Pleistocene Ice Age cycles in deep-sea sediments, they used the isotopic composition of sedimentary rocks to monitor environmental changes. But instead of measuring oxygen isotopes, they examined isotopes of carbon. These, like oxygen, can be fractionated when some process prefers one isotope over another. During photosynthesis, the plankton that live in seawater extract carbon from the oceans, preferentially taking up one of the isotopes of carbon, carbon-12, relative to the other, carbon-13. This leaves behind seawater enriched in carbon-13. The Harvard researchers found that in the Namibian rocks deposited just before the ice age began, the carbon isotopes are consistent with this normal situation. But in the glacial interval, and through hundreds of meters of the cap carbonates that overlay the glacial deposits, there is no longer evidence of an enrichment in carbon-13. The logical conclusion is that photosynthesis had ceased, that the ocean was effectively "dead," because it was frozen and had been that way through the several million years represented by the glacial and cap carbonate sediments. The CO_2 that would normally be consumed in photosynthesis instead built up in the atmosphere.

The beauty of an isotope signature of the kind described by Hoffman and his colleagues is that it can be checked in ocean-deposited rocks of the same age, no matter where they occur. They need not be associated directly with evidence for glaciation; if photosynthesis had been reduced to a low level in a frozen Snowball Earth ocean, the same change in carbon isotopes should be apparent everywhere. And for the most part, that seems to be the case. The same kinds of carbon isotope changes observed for the Namibian rocks are also seen in Northwestern Canada, in Spitzbergen in the Arctic Ocean, and in Australia. The widespread nature of the evidence indicates that regardless of the cause, the effects were global. Furthermore, the detailed analyses that have been conducted at these localities have shown that there were four, and possibly five, major glacial episodes between approximately 850 and 550 million years ago. Each one of these exhibits changes in the carbon isotope composition of seawater that are larger than anything seen elsewhere in the geologic record, and in each case both the onset and the termination of the glacial episode seem to have been very rapid. One interpretation is that each of these periods of glaciation was a separate Snowball Earth interval, with a completely frozen ocean.

Why should a series of Snowball Earth periods occur during the Late Proterozoic, and, as far as we know, at no other times in Earth history? Critics of the idea suggest that the very uniqueness of Snowball Earth episodes is good reason to doubt their existence. But there is circumstantial evidence that may explain why they occurred at this time and not at others. First, as we have already seen, the sun's energy output was lower than today. That alone is not a very compelling argument, however, because it was lower still prior to the Late Proterozoic, through a long period when there is no evidence of *any* significant glaciation on Earth. Secondly, as we have also seen, the continents at this time were clustered at low latitudes, a configuration that has not occurred since then. Oceans in tropical regions absorb and store solar energy and tend to warm the Earth, but when continents are present at low latitudes, they reflect sunlight, and this would have been especially true of the barren continents of the Late Proterozoic, which were not yet covered in vegetation. The net result would have been that considerably less solar energy than at present was retained by the Earth, causing global cooling. More speculatively, chemical weathering of the tropical and subtropical continents—one of the processes that remove CO_2 from the atmosphere—may have proceeded rapidly on the low-latitude landmasses, decreasing the Earth's ability to retain solar heat through a reduction of the greenhouse effect. Theoretical mathematical "models" of how the Earth would behave with low atmospheric CO_2, low solar energy input, and the continents located near the equator indicate that the oceans would cool and then begin to freeze at the poles. Because of positive feedback—as the ice began to extend from the poles toward the equator, more and more solar energy would be reflected back into space—the temperature would continue to decrease. The theoretical treatments also suggest that at some point, this would become a runaway process, and the entire ocean would freeze rapidly. This is consistent with the field evidence from Namibia and elsewhere, which indicates that the Late Proterozoic glacial episodes began very rapidly.

The Snowball Earth hypothesis seems to be consistent with virtually all of the Late Proterozoic geological features: evidence for simultaneous widespread glaciation, ice sheets at low latitudes, the sudden appearance in the geological record of banded iron formations, large changes in the carbon isotopes in seawater, and the juxtaposition of glacial deposits with the warm-water cap carbonates. Still, that does not mean it is correct in all details. There are alternative explanations for some of these features. One school of thought suggests that both the sudden warming at the end of glaciation and the carbon isotope changes in seawater could be explained by the release of a large amount of methane gas from frozen ground as continental glaciers began to wane. An important difference between this scenario and the Kirschvink hypothesis is that it would not require a frozen ocean. Methane is a powerful greenhouse gas, even more effective than carbon dioxide at trapping the sun's heat energy in the atmosphere. It is produced by bacterial decomposition of organic matter, and because it is a gas at most surface temperatures, it normally seeps slowly out of the ground into the atmosphere, where it is gradually destroyed by chemical reactions with other compounds. However, at low temperatures, it can be trapped underground in an icelike compound. Large amounts can be stored in the frozen ground of cold regions, as is the case in arctic permafrost today. Although the overall abundance of organic matter was much less during the Late Proterozoic compared to today, bacteria were widespread. Large quantities of methane could have been formed and stored during the long glacial intervals that characterized this time. Its rapid release could account for the "super greenhouse" episodes that apparently followed the Snowball periods. Methane produced by bacteria contains carbon with an isotropic composition consistent with the carbon isotope shifts observed in the Namibian sedimentary rocks, and those from other locations as well. Perhaps after all the Snowball Earth was not quite as frozen as some think. Perhaps there were tropical glaciers, but much of the low-latitude ocean remained open.

As fragmentary as the geological record is for the Late Proterozoic, things get even murkier farther back in the Earth's past. We really have no clear idea about what fraction of the preexisting rock of the Earth's crust—our only window back into the Earth's history—has been eroded away or caught up and transformed beyond recognition in mountain-building events. In spite of that, by piecing together information from all of the world's continents, it is possible to determine with a high degree of confidence that there were, at a minimum, two periods of worldwide glaciation, true ice ages, prior to the Snowball Earth episodes. One of these occurred between 2.2 and 2.4 billion years ago, the other at roughly 2.9 billion years ago. Equally important, however, is the fact that in spite of careful searching, there is no hint of an ice age in the vast stretch of time, approximately 1.4 billion years, between the Late Proterozoic events and the ice age that ended 2.2 billion years ago. That gap is 30 percent of the Earth's entire history, and its very existence must hold clues to the conditions necessary for glaciation. But it occurred so far back in the past that our knowledge of geological events then is limited. We do not have very reliable information, for example, about where the continents were situated on the globe, or how they moved about, or what their surface area was during this period.

We also have little information about the concentration of greenhouse gases in the atmosphere.

The evidence for one or more ice ages between 2.2 and 2.4 billion years ago is widespread. In central Canada and the United States there are glacially scratched and grooved bedrock exposures, finely layered varves with occasional large "dropstones" that are characteristic of glacial lakes, and tillitelike glacial sediments. Similar features are widespread in South Africa, and some also occur in northern Europe, although the European localities have been more heavily metamorphosed than those in North America or South Africa and are not as complete. There are also glacial sediments from this timespan in western Australia. Like the evidence of Late Proterozoic glaciation, that for the ice age 2.2 to 2.4 billion years ago is so far-flung that it suggests a global episode. The limited available magnetic data indicate that at least the South African glaciation, and perhaps some of the others, occurred at low latitudes. The Canadian deposits also include rocks that are remarkably similar to the Late Proterozoic cap carbonates. These features have led some to suggest that this early glaciation must also have been a Snowball Earth episode. But while such an idea is tantalizing, it is, at present, much less susceptible to rigorous testing than is the Late Proterozoic evidence. Dating such ancient events is subject to uncertainties that make it impossible to say with confidence that glaciation took place simultaneously in all localities—only that it occurred within the same overall time window. And whether the now-widespread locations where evidence for glaciation has been found were similarly distant 2.2 billion years ago, or whether they were once contiguous and were later dispersed by continental breakup, is also unknown.

The very earliest glaciation we know about on Earth, according to the best dating that is available, occurred between 2.9 and 3.0 billion years ago, in the Archean Eon, the most ancient of the subdivisions of geological time. Once again the evidence comes from South Africa. Spread across a distance of several hundred kilometers there are several outcrops of tillitelike rocks that contain pebbles and boulders clearly

Figure 19.   A glacially scratched and faceted boulder from the Earth's oldest known ice age. This sample comes from a tillite in South Africa that is estimated to be between 2.9 and 3.0 billion years old. Photograph courtesy Professor John Crowell, University of California, Santa Barbara.

scratched and faceted by glacial processes (figure 19). Similar deposits have also been observed deep underground in mines. There is little doubt that these record the workings of glaciers. The only question is, Was the glaciation a local phenomenon in a mountainous region, or is this evidence of an early ice age? With the scanty clues from South Africa, and so far no reports of glacial deposits in rocks of this age from elsewhere, the question is impossible to answer. There are, however, several discrete tillite layers, reminiscent of the multiple cycles of glaciation and deglaciation that are known from other, younger, ice ages. The existence of any decipherable record of glaciation at all from this distant time in the Earth's past is itself quite remarkable. The scratched and grooved rocks from South Africa could be the products of the Earth's very first ice age. On the other hand, there are also no compelling reasons to believe there were no earlier glacial periods. We may never know with certainty, because much older rocks are rare and for the most part have been buried, heated, and metamorphosed to the point where any signs of glacial activity would have been obliterated. There are, perhaps, a few secrets that Nature will not reveal.

CHAPTER NINE

Coring for the Details

The physical and chemical records of ancient ice ages stored in rocks—scratched and faceted boulders, glacial drift deposits hardened into coherent tillite, carbon isotopes, and various other signatures—allow us to trace glaciation back through almost three-quarters of our planet's history. However, we know almost nothing about the finer details of the frigid intervals that occurred before the Pleistocene Ice Age—what the actual temperatures were, whether there were repeated 100,000-year cycles of glaciation and deglaciation paced by the Earth's orbit, or how the climate varied across the globe. But over the past few decades, scientists have accumulated an amazingly detailed picture of all of these things, and more, for the most recent cycles of the Pleistocene Ice Age. Nearly all of this information has come from cores—cores of sediments from the seafloor, cores of ice from Antarctica and Greenland, and cores from lake beds and small mountain glaciers. Almost anything that accumulates or grows in a regular fashion has the potential to preserve a decipherable record of the environment. Even trees and coral heads have been cored, although in these cases the record usually does not extend very far into the past.

The information about ice age climates revealed by these types of records is not very comforting. The ice cores have produced especially dramatic results, because they allow an almost year-by-year reconstruction of climate history back through several glacial-interglacial cycles. They show that in the geologically recent past, wild but long-lasting swings in average temperature have occurred over periods as short as a few years. Climate shifts during the ice age were once believed to be slow and ponderous, but the new data show that sometimes they can be very rapid. Our distant ancestors lived through such periods, although we have no record of the effects on their lives. It is fairly obvious, however, that in the finely balanced twenty-first century world, drastic changes in the frequency and intensity of storms, in patterns of precipitation, or even just a significant change in average temperature, could wreak havoc with agriculture, trade, and transportation. The possibility that mankind's activities, such as adding the greenhouse gas CO_2 to the atmosphere through the burning of fossil fuels, might affect the natural changes in unknown ways adds great urgency to the need to understand what has happened in the past. This has not been lost on some of those likely to be affected. Small island nations in the Pacific have become ardent supporters of attempts to curb greenhouse gas emissions, basing their stance on research showing that rising sea level—which would cause them great damage—accompanies increased atmospheric CO_2 during interglacial periods. Some large insurance companies have been following climate change research with much interest, even to the extent of funding investigations into how global warming may affect the frequency of storms and floods. Self-interest is a potent motivator.

In chapter 7, I briefly discussed the crucial role that data from deep-sea cores played in confirming the connection between ice age climate and variations in the Earth's orbit around the sun, as originally proposed by Croll and Milankovitch. Coring the ocean floor is one of the most effective ways to unravel the Earth's past environment, and it is worth making a slight diversion here to explore the way in which this important technology has developed.

The concept is an old one. James Croll was probably the first person to suggest that clues to the history of ice age climates might be neatly stored away in the sediments of the ocean; he thought that the remains of land plants and animals would be washed into the sea and preserved, layer by layer, and that the climate changes of the ice ages would be reflected in the types of plants and animals buried in the sediments. Very little was known about ocean sediments when Croll wrote about ice age climates in the 1860s, but that was soon to change. A decade later a major expedition devoted to scientific exploration of the oceans was conducted aboard the vessel H.M.S. *Challenger*. The ship left England just before Christmas in 1872, embarking on a three-year voyage that traversed the world's oceans and made landfall on all continents, including Antarctica. Although the *Challenger* was part of the British Admiralty's fleet, and the voyage was subsidized by the Navy, the expedition itself was conceived by Charles Wyville Thompson, professor of natural history at the University of Edinburgh, and facilitated by the Royal Society of London. And while the British Navy undoubtedly benefited from its investment—the voyage revolutionized our understanding of the oceans—it was also a wonderful example of the creative use of military resources. The voyage of the *Challenger* is generally acknowledged to have been the first truly scientific oceanographic expedition, and in many ways, it marked the beginning of the modern science of oceanography.

The scientists on board the *Challenger* made meteorological observations, measured ocean currents, recorded seawater temperatures, collected biological specimens, and took water samples for chemical analysis. They also lowered dredges to the seafloor at regular intervals in order to collect bottom sediments—with the aid of some 270 kilometers of Italian hemp rope taken along for the purpose. One of the multitude of important discoveries they made was that in much of the deep ocean, far from the influence of continents, the sediment on the seafloor is a very fine grained ooze made up primarily of the shells and skeletons of tiny, surface-dwelling organisms—the plankton. Croll's idea that deep-ocean sediments would contain the fossils of plants and animals washed in from the continents was wrong. But the actual situation was even better, although this realization would come only gradually. Many different species comprise the plankton. Their life cycles are typically measured in days, and both the mix of species and the chemical composition of their shells respond rapidly to changes in the properties of the surface water. The continuous rain of dead plankton down to the seafloor, and their accumulation there, provides an unparalleled record of conditions at the ocean surface, stored in the layers of sediment. However, the dredges used on the *Challenger* expedition simply scooped up the muddy ooze from the bottom, mixing the layers together in the process and destroying the most important aspect of these sediments—the dimension of time that is preserved in them. The challenge for oceanographers would be to devise a way to sample the layers of deep-sea sediments without disturbing the original sequence of deposition.

As the science of oceanography evolved, the pressure to find a way to collect bottom sediments without distorting their vertical structure increased. On land it was routine to examine sequences of sedimentary rock layers in order to learn about the geologic history of a region. But geologists couldn't (at least in those early days) visit the deep seafloor for direct observations; sediment samples would have to be brought to the surface in their original configuration. Early attempts were predictably crude. Simply dropping an open pipe into the seafloor ooze would sometimes work, but there were multiple problems: How do you ensure that the pipe goes into the sediment vertically? How can you prevent the sediment from simply falling out of the pipe as you haul it up to the surface? In deep water, how do you know when the pipe is nearing the seafloor? How can you drive the pipe further into the sediments to obtain a longer core? With trial and error, and characteristic ingenuity, oceanographers gradually improved the design and use of coring apparatus. Fins and other stabilizing devices were added to the pipes to keep them upright as they fell through the sea like guided missiles. "Core catchers" were added, to prevent the sediment from falling out of the corers. Lead weights and sharpened, tapered nosecones were fitted to help the core barrels penetrate more deeply.

Most of the early devices were capable of retrieving cores that were only one or two meters long. Although sediments accumulate slowly over much of the ocean floor, often at rates of only a few centimeters every thousand years, two meters of core still doesn't represent a very large interval of geologic time—generally less than 100,000 years. Nevertheless, the early cores provided valuable insights. In the 1930s, a series of short cores recovered from the Atlantic showed that the mix of plankton species changes systematically with depth in the sediments. Based on what was then known about the conditions under which the different species grow, it was concluded that the changes probably reflected changes in surface water temperature—alternating periods of warmth and cold. This insight was a tantalizing foretaste of how sediment cores might shed light on ice age climates.

As the technology for retrieving sediment cores improved, so did the length of the cores, and therefore the timespan they represented. In the 1940s, a Swedish oceanographer named Börje Kullenberg made a breakthrough. He put a piston in the traditional core pipe and designed it so that the piston would be drawn upward as the pipe penetrated into the sediments, effectively pulling the sediment into the core barrel. Although there have been some modifications, his basic design is still in use today. The original version of the Kullenberg piston corer was deployed at regular intervals during the Swedish Deep Sea Expedition of 1947–49, which, like the *Challenger* voyage three-quarters of a century earlier, circumnavigated the globe. Kullenberg and the other scientists aboard the ship *Albatross* were ecstatic when the new device recovered cores up to 15 meters in length. In all they collected some 200 sediment cores; lined up end to end, they amounted to a ribbon of sediment one and a half kilometers long. Fulfilling Croll's dream, some of the cores they retrieved represented nearly a million years of sediment accumulation.

Aboard the *Albatross* in 1947 and 1948 was Gustaf Arrhenius, a young geologist from a prominent Swedish family. His grandfather, the chemist Svante Arrhenius, had been the first Swedish recipient of the Nobel Prize. Svante Arrhenius's interests were wide-ranging, and among his many accomplishments, he was the first to suggest that changes in the CO_2 content of the atmosphere could affect climate through carbon dioxide's ability to trap solar heat. Fittingly enough, it was his grandson Gustaf who discovered features in the long sediment cores taken with the Kullenberg corer that are directly linked to CO_2 changes in the ice age atmosphere. Examining cores from the Pacific Ocean, Gustaf Arrhenius found that the sediment layers were alternately rich and poor in calcium carbonate. The changes were quite regular, and Arrhenius concluded that they were somehow connected to glacial-interglacial climate cycles. Subsequent work has proven he was right about this—dating has shown that the timing of the sediment cycles corresponds well with glacier advance and retreat on land—although his interpretation, that they were due to variations in the intensity of ocean water circulation in the Pacific, turned out to be incorrect. The sediment cycles were later linked to large changes in atmospheric CO_2 that accompanied the Pleistocene glacial-interglacial changes. So interdependent are the atmosphere and ocean that changes in one invariably cause changes in the other, and this is particularly true of CO_2. As is the case with all gases, the amount of carbon dioxide that dissolves in seawater depends on its concentration in the atmosphere. And the amount of calcium carbonate that accumulates in deep-sea sediments in turn depends on just how much dissolved carbon dioxide there is in the oceans. Alternating layers of carbonate-rich and carbonate-poor deep-sea sediments would be expected if CO_2 in the atmosphere rose and fell through the glacial cycles. As we shall see, gas bubbles trapped in Greenland and Antarctic ice show that this is exactly what occurred.

The ten-meter cores collected during the Swedish Deep Sea Expedition were at the cutting edge of technology in their day, and they provided much information about ice age climate fluctuations, but they are a far cry from the several-kilometers-long cores that can be retrieved with present-day techniques. Today, drilling is the preferred method for obtaining long cores. Each metal core barrel that penetrates the seafloor has a plastic liner that fills with sediment as the drilling proceeds; when the cores are pulled up to the surface, the liners can be quickly and efficiently pulled from the barrel without distorting the layers. Typically, the cores are sliced in half lengthwise, and one portion put away as an archive, to be used, if at all, only when all other material is gone or new methods of analysis are developed.

The most ambitious and successful ocean sediment drilling program is the Ocean Drilling Project (ODP), the current incarnation of a program that has operated continuously since 1968. It utilizes a converted petroleum drillship, the *JOIDES Resolution,* which circles the globe taking cores of the seabed for scientific research. A massive drilling tower the height of a twenty-story building sits astride the *Resolution,* over a large hole—the "moon pool"—that penetrates the middle of the ship's hull, through which the drill string can be lowered directly into the sea. The *Resolution* is jammed with labs and computers and fitted with a bevy of thrusters, computer-controlled propellers that can keep it stationary at a drilling site even in heavy weather. Lowering the drilling rig to a target on the seafloor several kilometers below the ship is no mean feat, but together the *Resolution* and its predecessor in the project, the *Glomar Challenger,* have drilled into the ocean bottom at many hundreds of sites, in all the world's oceans, since the program began. Originally solely an American endeavor and later expanded to include international partners, the drilling program has been a notable success story among large-scale scientific projects. Drilling goes on around the clock. A typical "leg" of the *Resolution*'s continuing expedition lasts about two months and is usually focused on a particular scientific question in a specific part of the ocean. Between legs, there are short port calls for refueling, restocking, and exchange of scientific crews, which are usually international in character and eclectic in scientific background and expertise. One of the important goals of the drilling project is to unravel the details of the Pleistocene Ice Age through the sedimentary record of glacial-interglacial cycles. The availability of cores that extend back well beyond the inception of the current ice age also makes it possible to examine climate changes at even earlier times. Although the seafloor is much younger than most parts of the continents, its oldest sediments date back to about 200 million years (older seafloor than this has been destroyed in the great plate tectonic cycle of seafloor creation at the ocean ridges and consumption at ocean trenches). Having a 200-million-year window into climate and environmental change means that a considerable slice of Earth history can be examined in great detail.

Studies of ocean sediment cores, especially the comprehensive collection of materials from the ODP, show that the seafloor is far from the quiet, peaceful place it was once assumed to be—a sheltered library with page upon page of sediments neatly stacked up and waiting to be read. Instead, it has been found that bottom currents sweep through some parts of the seafloor, picking up sediments in one place and dumping them in another; that gravity works just as well at sea as it does on land, sending muddy avalanches down slopes after earthquakes or other disturbances; and that even in the deepest ocean there are living creatures that dig and burrow and churn up the sediments, blurring the layer-by-layer record. All of these processes complicate the interpretation of sediment core properties. But in spite of such difficulties, the amount of information that has been gleaned about the Earth's past climate is staggering. A good example is the record of ocean temperatures over the past 65 million years shown in figure 20. The data come from analyses of many different sediment cores, most of them from the Ocean Drilling Project. The temperature data are based on measurements of oxygen isotopes in the shells of small animals that live in the deep sea, and the timescale is largely based on interpretation of the magnetic properties of the cores. Both of these approaches were discussed in chapter 7. What is quite remarkable is that we are able to trace the trends in the Earth's climate over such long time periods with some confidence. The changing temperatures of the deep sea deduced from the sediment cores are believed to be a generalized reflection of

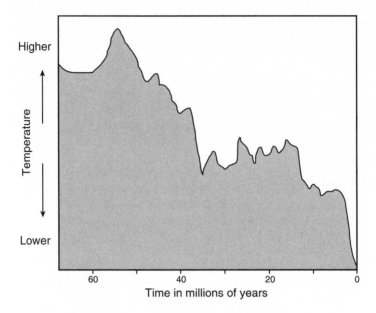

Figure 20. This qualitative graph, based on oxygen isotope analyses of fossils from deep-sea sediment cores, shows how ocean water temperatures—and presumably also average Earth surface temperatures—have varied over the past 60 million years. It is obvious that there has been a gradual decrease since about 55 million years ago, with especially sharp drops between 40 and 35 million years ago, and again during the past few million years. These are the times when glaciation began in the Antarctic and the Northern Hemisphere respectively.

changes in the average surface temperatures on our planet. It is clear from the figure that there has been a persistent temperature decrease from about 55 million years ago to the present, and also that there have been a few intervals with very rapid decreases—and some with increases—along the way.

Parameters such as the oxygen isotope composition of shells are generally referred to by geochemists as proxy indicators, or simply proxies, because they don't measure temperature or climate change directly. Rather, a proxy has to be translated into some environmental variable of interest through an understanding of how the proxy itself responds.

More and more proxies, many of them chemical and most of them measured in ocean sediments, are being added to the geochemists' arsenals. Nevertheless, oxygen isotopes remain one of the most valuable and informative proxies available to us. Even to a skeptic, the pattern of 100,000-year cycles in oxygen isotope compositions that characterizes the past million years or so, and that shows up again and again in cores from throughout the world's oceans (figure 16), is striking. It is hard to believe that the correspondence of these variations with the 100,000-year cycles of eccentricity of the Earth's orbit around the sun is purely coincidental. Nevertheless, as implied in chapter 7, interpretation of these changes is not entirely straightforward, because at least two separate factors influence the oxygen isotope composition of shells. One is the temperature at which the shells grew, something that is reasonably well understood, because it has been calibrated by both laboratory studies and measurements of samples from nature. The second is the oxygen isotopic composition of the seawater, which depends on the amount of glacial ice that existed on the continents when the shells grew. Why this should be so may not seem obvious at first, but the reason is quite straightforward. Evaporation from the oceans preferentially causes one of the oxygen isotopes to be enriched in the water vapor, leaving behind liquid water that is depleted in that same isotope. The oxygen isotopic composition of the ocean is therefore changed by this process. If the evaporated water is precipitated at high latitudes and ends up as glacial ice, the ocean's isotopic composition remains changed until the ice melts and the water again returns to the sea. The larger the volume of the ice on the continents, the bigger the shift in the ocean.

Fortunately, both colder temperatures and larger ice volumes change the oxygen isotopes in the same way, so the overall effect is to amplify the glacial-interglacial variability in the oxygen isotope record. This is fine for getting a general picture of the climate variations, but it would also be useful to disentangle the two effects. Recently, some exciting progress has been made on this problem. The approach has been to use a newly developed proxy that is independent of the oxygen isotope variations to determine past seawater temperature. With this knowledge, the expected temperature component can be subtracted from the oxygen data, leaving a residual record that should be due only to changes in the volume of continental ice sheets. Two very interesting insights have emerged from this procedure. First, it appears that ocean surface water in the tropical Pacific warmed up by 3–4°C during the most recent deglaciation, a much larger increase than earlier data had suggested (this also means that the tropical ocean had been much cooler during the glacial interval than earlier suspected). Secondly, the data suggest that there was a lag of two to three thousand years between the temperature increase and the decrease in ice volume. This is not very surprising—think about how long it takes for a bag of ice cubes to melt, or an old-fashioned refrigerator to defrost, even when the ambient temperature is far above freezing. But intuitive or not, this information could not have been gleaned from the oxygen isotope data alone—the simultaneous use of more than one proxy was the critical step. The same approach applied to the previous deglaciation, about 120,000 years ago, gives similar results. This new knowledge paints a detailed picture of *how* the glacial periods ended, not just when, and it suggests that warming in the tropics plays a crucial role in ushering in an interglacial interval.

And the story gets even better. Although ocean sediment cores have provided an array of proxies that can be used to track ice age climatic changes, cores of the ice itself have added an entirely new dimension. Just as the sediment cores do, they hold a number of proxies for ice age climate, but in addition they contain direct information about environments in the past. Entombed in the ice are air bubbles, samples of the ancient atmosphere that, in spite of their very small size, can be analyzed for many different constituents, even those present in trace amounts. Of special interest are the greenhouse gases that can trap the sun's energy and raise the planet's temperature. Measurements of these tiny time-capsule air bubbles have revealed that greenhouse gas concentrations in the atmosphere varied approximately in step with the cycles of glaciation and deglaciation. But deciding whether greenhouse gases are implicated as a cause of glacial-interglacial temperature variations or are simply a result requires very accurate timescales for both ice and ocean sediment cores, so that their respective records can be compared. Such accuracy is difficult to attain with conventional dating methods. But in recent years, an ingenious approach to this problem has been devised, based on the fact that the timing of the Earth's orbital cycles is very precisely known. If they really are the root cause of climate cycles, they can be used as a kind of template to examine the changes observed in deep-sea sediments or glacial ice. All that's required is to date one or more levels in the cores accurately and fix them relative to the orbital cycles—a procedure referred to as "tuning." Analyses of this type show that the CO_2 content of the atmosphere (based on ice-core measurements) and the Earth's surface temperature (based on oxygen isotopes in ocean sediments) both changed in sync with the 100,000-year eccentricity cycle of the Earth's orbit around the sun. These same analyses also show, as we saw earlier, that the volume of ice on the continents lags behind the temperature change—apparently by a few thousand years. The very close correspondence between temperature and CO_2 variations suggests that somehow carbon dioxide in the atmosphere is regulated by changes in the eccentricity of the Earth's orbit, and that it in turn regulates temperature. How this occurs is currently unknown. But these results once again confirm the Croll-Milankovitch theory—the regular orbital cycles act like a metronome, ticking out the rhythm of the planet's climate cycles.

As should be obvious by now, examination of ice cores is an important part of research into the Pleistocene Ice Age. But it is a fairly recent endeavor. The story of ice coring, especially in the Greenland ice sheet, has been nicely told in a recent book titled *The Two-Mile Time Machine* by Richard Alley, a scientist at Pennsylvania State University who has been deeply involved in that effort. In the following, I trace the impact of ice-core science on our understanding of ice ages, drawing on both Alley's book and other sources.

Coring a glacier, especially in the extreme climates of Greenland or the Antarctic, is no simple matter. Equipment has to be brought in, workers have to be housed and fed, and the cores, once collected, must be stored at temperatures well below freezing. If you hadn't thought about it very seriously—or even if you had—you might question why anyone would want to go to such effort just to collect a bit of ice. Part of the answer is the pure curiosity of mankind; it's like climbing a hill to see what's on the other side. A feature of glaciers that must surely have piqued the curiosity of many who observed them over the years, and that certainly played a part in the desire to core into them, is their visible layering. Like the layers of sedimentary rocks, the layers of ice in a glacier record the passing of time, and like the pages of a diary, each layer contains clues about what happened in the past. The information is cryptic, but with the right tools it can often be deciphered.

Early attempts to recover stratigraphic (i.e., layer-by-layer) information from ice date back to 1957, when the International Council of Scientific Unions, a body that coordinates international activities in science, launched the International Geophysical Year (IGY) to promote geophysical research on a global scale. By its nature, geophysics is international, and under the aegis of the IGY, scientific projects were conducted quite literally from pole to pole. Sixty-seven different countries had official roles, and, in spite of its name, IGY lasted for eighteen months. An important focus was the polar regions, which at that time were still poorly known scientifically. Both the Arctic and Antarctic held much interest for geophysicists, because they are the home of the north and south magnetic poles, the aurora borealis, and, of course, the largest remaining ice sheets of the Pleistocene glaciation. But the early ice-sampling attempts were quite crude. Hand-dug pits were made in the layered ice at places like Byrd Station, a U.S. encampment on the Antarctic ice sheet at 80° south latitude. These surface holes, however, only penetrated through very young ice. They provided samples and data for the topmost part of the Antarctic glaciers, but could not capitalize on the third dimension, the great thickness of the ice. The U.S.

Army Corps of Engineers already had a major laboratory devoted to "cold regions" research at the time of the IGY, and its chief scientist, a man named Henri Bader, pushed hard for scientific drilling of the ice. Although it didn't happen immediately, within a decade after the end of the IGY, his organization had cored deep into both the Greenland and the Antarctic ice sheets.

The earliest drilling was done on a ships-of-opportunity basis, in places where camps or scientific stations already existed. It produced a lot of new information about the polar ice caps, but it also soon became apparent that a more coherent strategy was required, and that the greatest rewards would come from drilling where the longest possible undisturbed cores could be obtained, or where the accumulation rate of snow was especially favorable for obtaining high-resolution records. Several countries, both individually and in partnership, made ambitious plans for polar drilling, and by the mid 1990s, several deep-drilling projects had been completed on the Greenland and Antarctic ice sheets, and others were active. Particularly important for research on the Pleistocene Ice Age have been cores from central Greenland and Vostok Station in the Antarctic. In Greenland, under formidable weather conditions, teams from Europe and the United States drilled a pair of holes just thirty kilometers apart, almost dead center in the continent and at the very summit of the ice cap. The rationale for this seemingly double effort was that independent analysis of the cores would provide a cross-check on the reliability of the data—and also, two cores would provide twice as much ice for critical analyses. Both projects reached a depth of about three kilometers, where the ice is more than 110,000 years old. Except for the very oldest sections of the cores near the bottom of the holes, where flow over the underlying rocks of the Earth's crust coupled with the great pressure of the overlying ice has distorted the layering, the agreement between the two drilling sites is amazingly good.

In January 1998, at the other end of the world in the Antarctic, ice that until very recently was the oldest yet drilled (more than 420,000 years old) was retrieved from a hole that reached 3,623 meters depth. Although ice cores have been collected at many sites in the Antarctic, including at the South Pole, this very long core from the Russian Vostok Antarctic Station has special importance, and has been heavily studied, because it reaches so far back into the Pleistocene Ice Age. It does so because annual snowfall is much less in the Antarctic than in Greenland, and a given thickness of ice therefore represents a far greater stretch of time than it does in Greenland. (In September 2003, a European consortium announced that ice they had recovered at another Antarctic site dates back to at least 750,000 years. Few data are yet available for this core at the time of writing.)

Ice drilling at Vostok (where the mean temperature is a chilly -55 °C) actually began in the 1970s and was for a long time purely a Russian endeavor, but later became a joint Russia-France-U.S. project. Based on remote sensing measurements from the surface, the ice continues for more than 100 meters beyond the depth reached by drilling. However, a decision was made to stop, because below the ice lies a large lake— Lake Vostok—that has been isolated from contact with the atmosphere for hundreds of thousands of years. Insulated from the frigid polar air above by three and a half kilometers of ice, and heated from below by the slow but steady escape of heat from the Earth, the lake remains unfrozen. It is thought that Lake Vostok may contain unique bacteria or other life that has long been isolated from the rest of the world's biosphere, and biologists and chemists are anxious to devise a contamination-free sampling plan. The last thing they wanted was to plunge a dirty drilling string through the ice into its undisturbed waters.

The ice cores have yielded a massive amount of information about the Earth's climate and environmental conditions through the few most recent glacial-interglacial cycles of the Pleistocene Ice Age, including data about the local temperatures at the drilling sites, the rates of snow accumulation, a record of distant volcanic eruptions (which are indicated by thin layers of very fine volcanic ash that is distributed globally through the atmosphere), and even the intensity of winds. The latter comes from analysis of the amount of ordinary dust and sea salt in the ice cores: the greater the wind intensity, the higher the content of these materials. And, of course, the ice also traps samples of the atmosphere as small bubbles, as we have already seen. The most recent part of the ice cores, especially for the time since the Industrial Revolution, also provides a very good record of how man's activities have affected the global environment. Anything that forms gaseous molecules, or attaches itself to the very tiny particles that are transported around the globe by winds, can end up being deposited in polar ice. Mercury, lead, and freon are just a few examples of substances that have been measured in glaciers and can be traced directly to industrial processes.

Some of the properties measured in the long Vostok core are shown in figure 21. A remarkable feature of such graphs is that almost all of the properties that have been measured in the ice show patterns similar to those of oxygen isotopes in deep-sea sediments over the same time period. The Vostok core is long enough to show four complete glacial-interglacial cycles, and the approximately 100,000-year periodicity is readily apparent. As with the deep-sea sediments, setting up a reliable timescale for the variations observed in the ice cores is critical for their interpretation. Here glaciers have one great advantage over ocean sediments: snow accumulates relatively quickly, and annual layers are usually easily discernible. Even deep in a glacier, where high pressure and the flow of ice have thinned them, the layers can often be distinguished. If annual layers can be counted—and if you can be certain that none are missing—then the cores can be dated very precisely and all of the interesting environmental proxy indicators can be placed accurately on a timescale. Also, because of the very high resolution provided by the annual layers, even events of quite short duration can be identified accurately. But counting layers one by one is a tedious process—can you imagine counting tens of thousands of layers without making an error? Automation has helped. It turns out that the electrical conductivity of the ice changes subtly with the seasons, because different amounts of various trace compounds from the atmosphere are incorporated into

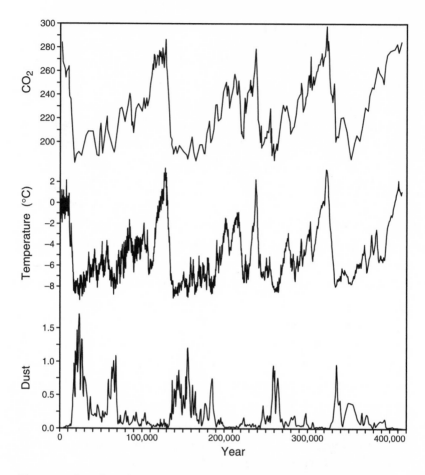

Figure 21.   Data from the Antarctic ice cores at Vostok Station show that temperature changes (relative to the present) calculated from isotopic data and atmospheric CO_2 content (in parts per million) from bubbles in the ice track each other very closely. Cold glacial periods had low CO_2, whereas the warm interglacials had much higher values. The amount of dust in the ice, on the other hand, is highest during the cold periods, signifying generally windier conditions. These graphs are based on data from a paper by J. R. Petit et al. in the journal *Nature,* June 3, 1999.

snow in summer and winter. Long sections of core can be scanned quickly for changes in conductivity, and the wiggles of the output interpreted in terms of yearly cycles. But the human eye and brain have remained primary tools for constructing ice-core timescales. Where cross-checks are possible, layer counting has proved to be very accurate. For example, the dates of quite a few large volcanic eruptions are well known from historical records that stretch back several thousand years. Fine-grained volcanic ash and chemicals such as sulfur dioxide that are spewed out in these eruptions are quickly distributed through the atmosphere and deposited on the icecaps as discrete layers. Dating by layer counting generally places these marker horizons in the ice cores within a few years of their actual occurrences. Uncertainties get larger for older parts of the ice cores, but in the Greenland cores, where layer counting has been very successful, dating appears to be accurate to a few percent over the past 100,000 years—a remarkable accomplishment. In Antarctic cores, too, layer counting gives correct ages for events that can be dated independently in other localities—for example, a particularly rapid change in global temperature that shows up in proxy records of different types in different places. Such agreement provides a high degree of confidence in the method.

The properties of the Vostok ice core shown in figure 21 illustrate some of the insights that have been gained into Pleistocene Ice Age climate from examination of polar ice. One, the close correspondence between temperature and the concentration of CO_2 in the atmosphere, we have already discussed. Both seawater temperatures, deduced from proxy measurements in sediments, and the local temperature at the Vostok drilling site—also based on a proxy, the isotopic composition of hydrogen in the ice—are closely correlated with carbon dioxide. The amount of dust in the Vostok core, a proxy for windiness, shows a strong anti-correlation with other parameters. It is obvious from the graph that the windiest times of the past 400,000 years, when most dust was deposited on the ice, occurred at the height of the glacial periods, when the temperature in the Antarctic—and presumably also globally—was at a minimum. This is consistent with other evidence that at the times of maximum ice cover, the global climate was cold, arid, and windy. In such an environment, the extent of desert areas increased, providing more dust, and grasslands expanded at the expense of forests. Extensive loess deposits formed in some parts of the world. Yet another feature that is apparent in figure 21, especially in the temperature and CO_2 plots, is the very sharp transition from cold to warm periods. This had already been noted in sediment cores, but it is even more striking in the Vostok data. Wally Broecker, an early investigator of the glaciation record in deep-sea cores, termed these quick shifts in temperature "rapid terminations," implying a swift end to glacial episodes after long periods of cold. The best estimate of the actual temperature change at the Vostok site during these transitions is about 12°C. Note, too, that the warm periods do not last very long—in fact, the present warm period is considerably longer than any other in the ice-core record—and that during each glacial period, the temperature gradually cools to a minimum value just before the next "rapid termination," giving the whole graph a saw-tooth appearance.

For completeness, there is one aspect of figure 21 that should be explained in more detail. It has to do with how measurements of the components of air bubbles correspond with other properties. Although both CO_2 and the temperature proxy were measured in the same ice core, at a given depth in the ice, these parameters do not record conditions at the same time. It turns out that an adjustment has to be made to the air bubble data in order to bring it into correspondence with the other properties. The reason is quite simple. As anyone who has walked in a snowfall knows, fresh snow is very light, because it is mostly air. But as more and more snow accumulates, pressure increases on the underlying layers, and air is squeezed out. On a glacier, as long as there are still spaces around the snow crystals, air in the snow will continue to exchange with the atmosphere, moving in and out with the winds and with changes in atmospheric pressure. But eventually the pressure of the overlying layers causes the snow crystals to grow into a continuous mass of ice, trapping whatever air remains as sealed bubbles and preventing it from further exchange with the atmosphere. That instant, when there is no longer communication with surface air, is the critical one for scientists attempting to decipher the gas bubble data. Depending on how quickly snow accumulates, this may happen at a depth in the glacier where the surrounding ice is anywhere from a few hundred to almost a thousand years old. In the latter case, a gas bubble analysis would refer to conditions that prevailed a thousand years after the temperature was recorded in ice at the same depth. Obviously, to make an accurate comparison of how various properties track one another, the timing must be known fairly accurately. Usually, it's possible to estimate the rate of snow accumulation from the thickness of annual layers in the ice cores, so that the uncertainty in the age of the air that's being analyzed can be minimized. Still, it should be realized that there may be small offsets in the records.

Up to this point, I have focused on the results from Greenland and Antarctic ice cores. That is where the remaining large ice caps are, and where the major drilling efforts have been mounted. But a few people have recognized that mountain glaciers, virtually all of them small remnants of much larger ice fields that existed at the height of the most recent glacial period, may also have interesting stories to tell. Because these smaller glaciers are widespread across the Earth, they also may hold clues about the global reach of the ice age climate. In particular, those that still exist at tropical or subtropical latitudes are often the only available storehouses of information about ice age climate away from the harsh environment of the polar regions.

Many scientists credit one man, a glaciologist at Ohio State University named Lonnie Thompson, with almost single-handedly inventing and shaping the study of ice cores from small mountain glaciers. In recent years, Thompson's decades-long contributions to climate science have been thrust into the limelight as he won a series of awards. These include the Heineken Prize of the Royal Netherlands Academy of Arts and Sciences, and, together with his wife Ellen Mosley-Thompson (also a glaciologist), the prestigious Common Wealth Award for Science and Invention. Commenting on his work, Thompson noted that the glaciers he studies are like canaries in a coal mine—sensitive indicators of change. Many are fast disappearing. It was the Thompsons who predicted that the Snows of Kilimanjaro—

mentioned earlier in this book, and one of a few equatorial glaciers—will completely disappear by 2020. Less well-known glaciers in South America, Asia, and North America are vanishing just as quickly.

Lonnie Thompson's early ambition was to be a coal geologist—he came from West Virginia, where coal mining is a major industry. But his horizons broadened as he worked toward a Ph.D. at Ohio State University. The university had been a center for data from polar regions collected during the IGY, and soon Thompson and his wife were both working on ice-coring projects at the university's Institute of Polar Studies (now the Byrd Polar Research Center). But research on Antarctic and Greenland ice was a new and very hot topic, and Thompson began to look for a niche not already filled by senior scientists and ambitious younger colleagues. While still a graduate student, he made an expedition to explore the five-kilometer-high Quelccaya glacier in the Peruvian Andes. That trip, physically demanding though it was, convinced him that drilling on the glacier was possible and would provide information that couldn't be had from the cores in Greenland or the Antarctic. Thompson wrote a proposal to the U.S. National Science Foundation (NSF) to mount a full-scale drilling program along the lines of those that were already operating in the polar regions. But his request was turned down—the consensus was that it just wasn't feasible to drill at such high altitudes. Thompson was not deterred, and managed to carry out annual summer research on the Peruvian glacier from 1976 onwards. By 1979, he had won over the NSF and secured money for drilling. Although his first attempt was a failure—flying the drilling rig to the glacier proved too dangerous—he eventually succeeded, in 1983, using people and pack animals to bring in the equipment and supplies. Thompson and his crew drilled through the glacier to the underlying rock—a short distance by the standards of polar ice cores, only some 160 meters—and obtained a well-preserved sequence of annual layers that extended back 1,500 years. It was the first core through a tropical glacier. Even so, some in the science community were not impressed, because the timespan was short. But for those interested in tropical climate, especially South American climate, it was a bonanza. Here, in great detail, was a record of environmental change. Variations in average temperature, wet periods, dry periods, even volcanic eruptions were all recorded in the cores. Because all of these would have affected the peoples of the region, archeologists could begin to make connections between climate and what they knew about the history of various societies in the area. Thompson's persistence had paid off.

From that first core in Peru, Thompson never looked back. In the intervening years, he and his colleagues have brought back high-altitude ice cores from China, Bolivia, Tibet, Alaska, and Mt. Kilimanjaro itself. But there was a price to pay in physical hardship—in the tropics and subtropics, permanent glaciers can only exist at extreme altitudes, and the logistics of living and drilling at such heights are equally extreme.

Thompson's record was in Tibet, where, in 1997, he and his crew drilled into a glacier at 7,200 meters above sea level—an amazing feat. Even in the cold of that high elevation, there is the question of whether there might be gaps in the record—warm years when the winter snowfall melted away in the summer sun. In the small mountain glaciers, annual layers in the ice are not nearly as pronounced as they are in the polar ice caps, and dating is difficult. Often a few marker layers—dust from a volcanic eruption that has been dated elsewhere, or a change in some chemical parameter that is recognized in other cores—are the only reliable sources of age information. Still, it is clear that some of the mountain glaciers contain ice that dates back well into the last glacial period. A core from Bolivia penetrated ice that appears to be at least 25,000 years old, and, although there is controversy about the exact ages, Thompson believes that some of the Himalayan glaciers contain ice that is hundreds of thousands of years old.

Perhaps the greatest value of the ice cores from mountain glaciers is that they provide an almost global coverage. Early workers on glacial climates assumed that the tropics probably enjoyed a fairly stable climate even while high latitudes suffered through the extremes of the Pleistocene glacial-interglacial cycles. Proxies for temperature in deep-sea cores from the tropics seemed to be in agreement with this conclusion, and in any case, it had been well known since Milankovitch made his calculations that variations in the incident solar energy in the tropics during orbital cycles are much smaller than they are nearer the poles. The ice-core data from Greenland and the Antarctic provided accurate information about the large temperature shifts at high latitudes. But until the work of Thompson—now joined by several other groups drilling mountain glaciers—there were no such data for regions nearer the equator. Studies of the low-latitude glaciers, combined with new approaches to studying ocean sediment cores from the tropics and large-scale computer simulations of the Earth's climate, have now shown that equatorial regions play a more important role in climate change than was believed earlier. This seems to be true of changes that range all the way from the glacial-interglacial cycles of the ice age to the four- or five-year timescales that characterize recurrent El Niño events.

The ice cores from low-latitude glaciers provide a record of changing local climate that stretches far back into the past, but the work has also focused attention on the present-day fate of these ice caps. Most of them are melting—fast. Lonnie Thompson and his colleagues, after completing a drilling project on Mt. Kilimanjaro that recovered ice cores spanning 11,700 years of climate history for tropical Africa, made headlines around the globe when they predicted that the famous glacier would be gone within a few decades if the current climate persists. They noted that during the twentieth century, the extent of Kilimanjaro's ice cap had decreased by 80 percent (some of this decrease may be due to local conditions rather than global warming). Large-scale shrinkage has been documented for other low-latitude glaciers as well. This melting back is not just a regrettable natural phenomenon, or loss of another tourist attraction for curious sightseers. Nearly three-quarters of the world's population lives in the tropics. Some depend on glacial meltwater for at least part of the year; it may not be a reliable source for very much longer.

CHAPTER TEN

Ice Ages, Climate, and Evolution

Mrs. Antrobus: What about the cold weather?

Telegraph boy: Of course I don't know anything . . . but they say
there's a wall of ice moving down from the North, that's what they
say. We can't get Boston by telegraph, and they're burning pianos in
Hartford.

. . . it moves everything in front of it, churches and post offices and
city halls.

Thornton Wilder, *The Skin of Our Teeth*

Weather is always a topic of conversation, and even today's city
dwellers, quite insulated from the natural world, tune in to the Weather
Channel to learn the latest about weather in their region or across the
globe. Climate is simply average weather on a long timescale. Thornton
Wilder's play *The Skin of Our Teeth* deliberately mixes geological periods—modern Americans, dinosaurs, and an ice age exist simultaneously—and weather, as often in literature, is a metaphor for crisis and
conflict. Some anthropologists and biologists think that in real life, it is
much more than a metaphor: that climate—especially the climate of the
Pleistocene Ice Age—has had a direct effect on human evolution. The
arguments are circumstantial and have to do mainly with timing. But
more complete fossil records for other species show strong links

between evolution, extinction, and the glacial-interglacial climate swings of the Pleistocene Ice Age. And much farther back in the Earth's history, ice ages may also have had an important influence on the way life evolved on our planet.

The evolution of modern human beings can be traced back through a number of species, collectively known as the hominids, to a common ancestor with the chimpanzees, our closest living relatives among the primates, some five or six million years ago. As far as we can tell, chimpanzees and other members of the family of "great apes" have not changed radically from that distant ancestor. We, on the other hand, have changed a lot. The interesting questions are, Why? and, How? Clues to these questions are hard to come by, and we may never know the answers with complete certainty. But the timing at least suggests an interesting connection with climate and the Pleistocene Ice Age. For several million years after our common ancestor, hominids evolved slowly. They developed the ability to move around on the ground with an upright posture, although their body structure suggests that they were still expert tree climbers. Then, right around the time when the Earth's average temperature plunged downward at the beginning of the Pleistocene Ice Age—about three million years ago—the rate of change accelerated drastically. Hominids quickly evolved away from their apelike ancestors, developing increasingly sophisticated tools and weapons, hunting, planning, complex language, and eventually agriculture, writing, airplanes, and computers. During that time there was an increase in brain size by more than a factor of three in less than three million years, a breathtakingly rapid change compared to the normal course of evolution. That change took place entirely within the Pleistocene Ice Age. Is there a cause-and-effect relationship? The detailed climate records from deep-sea sediments and polar ice cores discussed in chapter 9 have provided evidence that is, at the very least, suggestive of a link. This raises the question, Would we be here at all were it not for the Pleistocene Ice Age? Exploring that question can be a bit unsettling. Based on the evidence that is currently available, it seems possible that we are here, not because we won out over other species in some survival-of-the-fittest battle, or because of an inevitable march of evolution toward higher intelligence, but because of the fickle ice age climate of Africa. If this turns out to be true, the evolution of our species really was a roll of the dice, a chance outcome that could not easily have been predicted.

To investigate this premise requires a closer look at several aspects of the problem: What is it that really makes us human, and when in our history did these traits evolve? And, what was the ice age climate really like as these characteristics arose, and can a logical connection be made? Quite a few biologists, paleontologists, and anthropologists have examined these questions, especially in the recent past. Their main impetus has been the unequivocal coincidence in timing between the start of the Pleistocene Ice Age and the beginning of rapid change toward modern humans among the hominids. But this single observation could indeed be coincidence. After all, most of the fossil evidence for early hominid evolution comes from tropical Africa, a region far removed from the large-scale continental ice sheets of the polar regions. The climate cooled in tropical Africa as the ice sheets expanded, but temperatures remained moderate. And at higher latitudes, where temperature changes were much more extreme, many species dealt with the ice age climate swings, not through rapid evolutionary change, but simply by migration to regions of more equable climate. One would not, a priori, expect annual temperature changes of a few degrees Celsius to have affected early hominids very severely.

But let's look first at those characteristics that are generally agreed to distinguish us from all other creatures on this Earth. First is an upright posture, the ability to walk on two feet rather than four. The second is more important: intelligence that is much more advanced than that of even our closest relatives among the primates. It is that developing intelligence, linked to the rapid increase in brain size observed among the fossil hominids, that led to highly coordinated body movement, sophisticated tool-making abilities, and the emergence of complex language as a means of communication. Yet another characteristic of humans is that we mature very slowly compared to most animals. A human infant is completely helpless at birth, wholly dependent on its parents for survival, incapable even of locomotion for a year or more.

In the classical view of evolution as a gradual process that winnows out undesirable characteristics and preserves useful ones, none of the physical traits just mentioned—bipedalism, a larger brain, slow maturation—would seem to be especially advantageous. Walking or running on two feet, especially with the waddling gait that must have characterized early hominids with hip structures that had not yet fully adapted to upright posture, would not be a very effective way either to escape fleet-footed predators or to chase fleeing prey. Bigger and bigger heads at birth could mean death for both mother and child—a very effective roadblock to proliferation of any species, especially one with a small number of offspring. Helpless infants are not an efficient way to propagate a species either—they are unable to escape predators on their own. And yet our species evolved with those characteristics. Several scientists, notably the paleontologist Steven Stanley and the biologist-neurophysiologist William Calvin, have developed plausible scenarios for the evolution of *Homo sapiens* with these traits, scenarios that explicitly make a connection with ice age climate. A good deal of what follows is drawn from their writings.

Darwin was the first to recognize that humans had evolved from primates much like the present-day apes of Africa, which he thus viewed as the probable cradle of our species. His ideas about our evolution from apelike predecessors, now known from fossil evidence to be wrong in detail, are often portrayed in popular culture by the frequently reproduced cartoon of an ape gradually achieving upright posture and ending with a modern human in a suit carrying a briefcase. In reality, even our distant ancestors, the australopithecines, had upright posture. This is evident from their fossilized skeletons, and was dramatically confirmed by the discovery in Tanzania of a set of fossil footprints preserved in volcanic ash. A pair of barefooted, bipedal australopithecines strolled across a bed of soft volcanic ash that was wet from recent rain. The ash, like wet sand on a beach, made perfect casts of their feet. Their footprints—and those of other, smaller animals that passed the same way—were perfectly preserved when another layer of ash was deposited over them. We are fortunate that the eastern African home of these early hominids is an area of active volcanism, because volcanic ash layers are ideal for dating. They are often spread over a wide region essentially instantaneously, and they contain freshly crystallized minerals that can be separated and dated. The Tanzanian ash has been dated to 3.2 million years ago, and the footprints it contains gives us a tiny, random glimpse into the lives of our ancestors. That something as ephemeral as a footprint should be preserved over such a great span of time might seem surprising. But in fact fossil footprints, although not common, are found throughout the geological record. Dinosaurs, especially, are well represented in this way—and some of their preserved footprints are more than 100 million years old.

Australopithecines like the ones that left their tracks in Tanzania had small brains, similar to those of modern-day apes. As far as can be gleaned from the fairly sparse fossil record that remains, *their* ancestors had also had similar-sized brains for many millions of years. Things were going pretty well for the australopithecines. They were largely vegetarian foragers; the African forests provided both food and shelter from terrestrial predators such as the large cats. Although they could walk upright, australopithecines were also expert climbers. They had long arms that they could wrap around a tree trunk to help them shinny upwards like a telephone repairman. They were, in evolutionary terms, in a period of stasis. Small differences evolved among them, but they were on the whole in balance with their environment, with little pressure for radical change. But looming ahead was a crisis that would very quickly put an end to this comfortable situation.

In the 1950s, the eminent biologist Ernst Mayr developed the idea that new species often evolve when some small subset of an existing population becomes isolated from the rest of their species. In such circumstances, change can occur rapidly, especially if some trait or physical characteristic is favored reproductively—in other words, if individuals with those characteristics are more successful breeders. Eventually, the isolated population evolves so far from its ancestors that it can no longer interbreed with the parent population and a new species has been born. Sometimes the new species dies out, but often the newcomers eclipse their parents.

Mayr's work on the theory of speciation was founded on observations of modern species. Paleontologists, who study evolution by examining fossils, didn't pay much attention to Mayr's ideas until the 1970s, when Niles Eldredge and Stephen Jay Gould proposed their theory of punctuated equilibrium. In many respects, this was an extension to the fossil record of Mayr's and others' work on modern speciation. It envisioned abrupt appearances and disappearances of fossil species, with long intervening periods of little or no change. The probability of finding any transitional forms—Mayr's small, isolated populations—would be very small, because of the incompleteness of the fossil record. In spite of the fact that punctuated equilibrium explained many aspects of the fossil record, it had numerous critics and prompted a great deal of debate. Most paleontologists clung to the idea that evolution was gradual, and punctuated equilibrium proposed just the opposite. Many also thought of Darwin as the champion of gradual evolution and were wary of any theory that seemed to contradict his ideas. However, a careful examination of Darwin's writings shows that he didn't really describe all evolution as a slow and steady process. Instead, he recognized the importance of geographical distribution and isolation, and he realized that small populations were more amenable to rapid change than large ones. But even if they accepted that there had been periods of rapid evolution, one aspect of the punctuated equilibrium theory disturbed many workers: that the periods of equilibrium are often very long. They expected there to be a component of gradual evolution even when there was no significant external pressure for change. Yet many fossil species exist virtually unchanged for millions of years and then suddenly go extinct.

Something like the punctuated equilibrium model, shaped by climate changes of the Pleistocene Ice Age, may apply to human evolution. Unfortunately, the fossil record of hominids is very far from complete, which makes piecing together the details quite difficult. In contrast to ocean-dwelling organisms, those that live on land are rarely well preserved. On the seafloor, the steady rain of sediment buries and protects fossils; on land, bones get scattered by predators or scavengers, carried away by floods, or even blown about in windstorms. In spite of intensive collection efforts, there may be entire hominid species not yet represented in our fossil collections. Most of the specimens that exist come from special environments—caves, lakeshore settings where fluctuating water levels periodically buried skeletons in a protective blanket of mud, and regions where frequent volcanic eruptions laid down ash layers that buried and preserved hominid remains. And yet, in spite of the fragmentary evidence, the history of hominid evolution looks very much as though it follows the isolation-then-sudden-change model. The australopithecines, who, as we have seen, seemed to be comfortably successful in Africa, disappeared quite abruptly, with only a short overlap with our own genus, *Homo.* Another *Australopithecus*-like hominid, called *Paranthropus,* coexisted for more than a million years, then became extinct. Several species of *Homo* that we know about appeared and disappeared over the past 2.5 million years before *Homo sapiens* finally arrived on the scene, probably between 100,000 and 200,000 years ago.

For those attempting to establish a link between human evolution and the Pleistocene Ice Age, it is the overall climate and its effect on the environment, not just temperature, that is the key. The oxygen isotope data from deep-sea cores, discussed in the previous chapter, show that seawater temperatures, and by implication surface temperatures in general, have been on a downward trend since about sixty million years ago. Especially rapid decreases occur around thirty-five million years ago, when glaciers started to accumulate on the Antarctic continent, and also beginning about three or four million years ago. By then, ice caps were forming in northern Europe, Greenland, and North America. By two and a half million years ago, the Pleistocene Ice Age was in full swing in the Northern Hemisphere, with ice sheets advancing and retreating over the continents at regular intervals. Both ice and sediment cores exhibit high dust contents during the glacial periods of the Pleistocene Ice Age, indicating that the cold periods were not only windy but also dry. As global cooling proceeded, the main result in tropical Africa was that forests, the natural habitat of the hominids, shrank, primarily due to aridity, not the cooler temperatures. The forests probably grew back a bit during the warmer and wetter interglacial periods, but the overall trend was toward much less tree cover; in place of the forests, grasslands expanded. It is this change in vegetation, not the actual temperature variations, that seems to be the key to human evolution.

Before the Pleistocene Ice Age began, the widespread grasslands of central and eastern Africa did not exist. There were no great herds of antelope and other ungulates that migrated with the seasons, following the rains to greener pastures. There were similar animals—fossils show, for example, that there were precursors of the modern antelope. However, they were not adapted to the grass of the savanna that their descendents feed on today; instead, they ate the tender leaves of shrubs and small trees in open woodlands. It is quite clear that the environmental transformation engendered by the ice age forced evolutionary change among the antelope. The question is, How did it affect the hominids?

We saw from the previous chapter that temperature changes during the ice age are sometimes very abrupt, especially the so-called rapid terminations that characterize the switch from cold to warmer temperatures as recorded in the ice-core data. As we have also seen, the glacial-interglacial cycles of temperature change occur regularly on the 100,000-year timescale of the eccentricity of the Earth's orbit. But before about 800,000 years ago, for reasons not fully understood, the cycles of the ice age were shorter—closer to 40,000 years—and the glacial intervals may not have been quite as severe. (Note that 40,000 years is the approximate cycle time for the tilt of the Earth's rotation axis, the orbital parameter that Milankovitch thought would be especially important. In spite of the different cycle length, it seems that there was still an astronomical control of glaciation and deglaciation.) Earth was thus seesawing between cold and warm, dry and less dry, more than twice as fast during this earlier phase of the ice age. Such rapid climate switches may have engendered what William Calvin refers to as "boom and bust" cycles among early hominid populations (and later ones, too). In this scenario, the cool and dry periods were stressful. Only those populations with specific adaptive characteristics were successful in the changed environment, and in the boom times, when life was much easier, these surviving populations flourished and expanded rapidly. They also produced offspring with a whole new range of biological variations. Because the living was (relatively) easy in the boom times, these variations could persist even if they were neutral or even slightly negative as far as survival was concerned. Occasionally, these characteristics might be quite unexpectedly helpful when the going suddenly got tough as the next cold and dry glacial cycle began.

Our australopithecine ancestors were predominantly woodland dwellers. Probably they frequented the open woodlands on the margins of denser forests, like the predecessors of the present-day antelope mentioned above. Lacking natural protection and unable to travel quickly through the forest canopy like their close relatives the chimpanzees, they were relatively easy targets for the large carnivorous predators that abounded at that time. They did climb trees for safety and probably slept in them at night to avoid danger on the ground, but the shrinking and fragmenting woodlands of the building ice age meant that their population began to shrink and fragment too. It was a classic scenario for the punctuated equilibrium model of evolution. Steven Stanley refers to the events that led to the human lineage as the "terrestrial imperative." Our ancestors were literally forced to come down out of the trees to survive. They did not suddenly become bipedal because of the climate; they already were. But the vagaries of the ice age climate eventually forced them to abandon the arboreal part of their existence and become full-time ground-dwellers.

Because the hominid fossil record is quite sparse, it is not currently possible—and may never be possible—to trace out exactly what happened among the hominids between two and three million years ago. But we do know that *Australopithecus* had a small brain, and the very first species of *Homo* that we know about had a significantly larger one. All who deal with human evolution agree that this has to do with the slowdown of the rate at which *Homo* matured, a process often referred to as juvenilization. Juvenilization is by no means restricted to humans. Simply put, it is a kind of backing down of development, so that adults end up having features that, far back in the history of the species—or in a predecessor species—were juvenile characteristics. One result in humans is that newborn infants are helpless; another is that, unlike those of other primates, our brains continue to grow at a rapid rate after birth, approximately doubling in size during the first year. Steven Stanley has argued that juvenilization, and especially the resulting bigger brain of *Homo* compared to *Australopithecus,* was a direct result of the cool, dry ice age climate of Africa that forced our ancestors to abandon tree-climbing and become firmly terrestrial. He suggests that australopithecine infants must have been fully capable of clinging onto their mothers as they clambered up trees for protection; in contrast, ground-dwelling *Homo* youngsters were much less tactile, and their mothers needed to have hands and arms free to carry them. The species would have had the luxury of slow development, no longer requiring the infant survival skills of its predecessors. And when they weren't occupied with babies, hands freed full-time from their tree-climbing duties could also be employed for communication, tool-making, and throwing—all activities that would have required new developments in the brain, and an increase in its size.

Thus there are plausible arguments that link the beginnings of our genus with the onset of the Pleistocene Ice Age. But the story does not end with the "terrestrial imperative" and the advent of *Homo.* The entire history of humankind has been played out against the backdrop of the seesaw advances and retreats of massive continental ice sheets, and the worldwide climate effects that accompanied them. The same kinds of evolutionary principles that may have led to *Homo*—isolation of populations due to environmental changes, boom-and-bust cycles, and adaptation to new environments—almost certainly continued to play a role in our evolution, virtually up to the present time. Again the sparseness of the fossil record limits our ability to know all of the details of the climate-evolution links. But again there are many coincidences between the available fossil evidence and the details of the global climate record deduced from ice and sediment cores. In aggregate, they suggest a significant connection between ice age climate change and our evolution as a species.

Most of our information about climate during the early part of *Homo*'s existence comes from deep-sea sediment cores; the ice cores, which have the advantage that they provide climate details at higher resolution, are so far restricted to about the past 500,000 years. But one of the most important results of the ice-core research is the discovery that the well-documented 100,000-year cycles of the ice age themselves comprise a series of regular climate variations of shorter durations: there are cycles within cycles within cycles. Science reporters writing about this research have described abrupt temperature changes of these cycles as "flip-flops" and have characterized the ice age climate as "chattering" or "jittering." In spite of those adjectives, however, the changes are not entirely chaotic. There seems to be a rough regularity on a variety of timescales that extends all the way down to cycles of only a few years.

The realization that climate swings in the past have often been very rapid has spawned a whole new subdiscipline in climate research: abrupt climate change. It has also generated a lot of concern, because of the implications of unexpected climate shifts for society—in 2002, the National Research Council of the National Academy of Sciences in the United States issued a detailed report on the subject that was subtitled *Inevitable Surprises*. One of the most discussed of those surprises is a

Figure 22. *Dryas* flowers brighten the summer tundra in the Canadian arctic. About 12,800 years ago, *Dryas* suddenly appeared in Europe after a long absence, marking the sudden drop in temperatures that led to the cold interval called the Younger Dryas.

relatively cold period called the "Younger Dryas." Geologists and climatologists had known about this interval for some time, but the speed with which it began and ended were only realized when high-resolution data from ice cores became available. The Younger Dryas started quite suddenly 12,800 years ago, lasted for about 1,200 years, and then ended quite abruptly (there is an "Older Dryas" too, a much shorter cold period that occurred about one thousand years earlier). The name given to this cold interval comes from a species of *Dryas,* a small flowering plant that thrives today in the arctic tundra and alpine regions (figure 22); its leaves and pollen, along with those of other arctic plants, suddenly appear abundantly in European sediments dated near 12,800 years. Work in other parts of the world shows convincingly that the Younger Dryas event was not confined to Europe, however—it was a global phenomenon. The Younger Dryas cold interval was a sudden reversal of the generally rising temperatures that followed the peak of the most recent glacial episode some 20,000 years ago. The Greenland ice cores indicate that temperatures dropped abruptly, within a few decades, at the beginning of the Younger Dryas, then rose even more quickly—apparently in less than a decade—at its end (figure 23). Depending on location, temperatures seem to have been anywhere from 2–3° to 7–8°C cooler during the Younger Dryas than they were immediately before it began. The ice cores also contain much more dust during this interval, suggesting widespread arid and windy conditions and bolstering evidence from other localities that this was not just a local cooling in Greenland.

The Younger Dryas lasted for thirty or forty generations of our *Homo sapiens* ancestors living twelve thousand years ago. Both its inception and its end occurred within a single generation. Because of their rapidity, both of these changes would have caused sudden disruptions in the ecosystems upon which our ancestors relied for food and shelter, and those generations having to deal with the abrupt changes would have been severely stressed—especially those living at high latitudes. They would have had to deal not only with a large shift in average temperature, but also with changes in the pattern of precipitation, and the availability of familiar plant and animal species. It is plausible to infer that events like the Younger Dryas caused population fragmentation and the kind of boom-and-bust scenarios that William Calvin and others believe have affected human evolution.

The Younger Dryas occurred relatively recently in the history of the genus *Homo,* and there is no obvious aspect of human evolution with which it can be associated. It is much more likely that prolonged series of such rapid climate shifts were influential in our evolution. We know from ocean sediments that 40,000-year climate cycles were affecting the Earth for roughly the first two million years of *Homo*'s existence, and by analogy with the more recent past documented in the ice cores, there were probably also shorter, rapid cycles superimposed on those longer periods. For the most part, these shorter cycles do not show up in

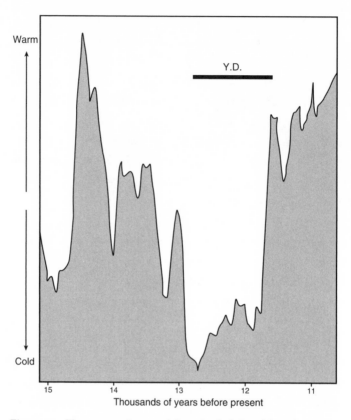

Figure 23.   Temperature in central Greenland, deduced from isotopes in Greenland ice cores, dropped suddenly 12,800 years ago at the beginning of the Younger Dryas interval, and stayed low for about 1,200 years. At the end of this period, the temperature increased by about 8°C in a decade or less. The vertical scale in this figure, and in figures 24 and 25, is highly exaggerated. All three graphs are based on data from a paper by P.M. Grootes and M. Stuiver, which appeared in the *Journal of Geophysical Research* 102 (1997): 26455.

the deep-sea cores. Sediments typically accumulate too slowly and are churned up sufficiently by worms and other bottom-dwelling organisms that even thousand-year events like the Younger Dryas are usually not discernible. Records of much shorter events, such as the decade-long temperature changes that can be resolved in the annual layers of an ice core, are completely indecipherable. Nevertheless, there are a few unique records, such as ancient lake sediments, that suggest abrupt climate changes like that of the Younger Dryas have occurred throughout the history of our genus. Rapid climate change and its "inevitable surprises," including population fragmentation, forced migration, and boom-and-bust cycles, have probably been a fact of life since humans appeared on planet Earth.

At about the same time that species of *Homo* appear in the fossil record, so do crude stone tools. Toolmaking is one of the traits that distinguish us from our australopithecine ancestors, and it requires a brain that can make connections among events and is capable at some level of advance planning—the tools were made for a purpose. The earliest tools date to about 2.4 million years ago and are thought to have been made by an early human species called *Homo rudolfensis,* which already had a brain almost twice as large as that of *Australopithecus* and (based on studies of its teeth) a much more varied diet. The tools were probably used for hunting and/or butchering animals. How and why did this species learn to make and use tools? Most of the ideas about human evolution from the time of *Homo rudolfensis* up almost to the present are in the realm of working hypotheses—they are based on available fossils and climate records, and they are designed to be tested as new evidence is discovered, and, when necessary, discarded and supplanted by a new hypothesis. One of the current working hypotheses is that as the ice age climate of Africa became progressively cooler and drier, but also cycled rapidly between wetter and more arid intervals, early humans learned to be effective hunters. Food of the shrinking forests became less bountiful; they were forced to turn to creatures of the grasslands to survive. They became meat eaters, and in the process they had to acquire a whole range of new skills that would have required new mental capabilities. Planning, cooperation, and throwing skills, the latter requiring a high degree of brain-body coordination, would have been essential. Think of a skilled pitcher throwing a fastball precisely to a catcher's glove, or of Wayne Gretsky blasting an accurate slap-shot past a goalie. None of our ape relatives is capable of such exquisite coordination; it requires a larger, human, brain.

One rather enigmatic stone tool that has been directly connected (in another working hypothesis) to ice age climate is an object called the Acheulean hand axe. It first appeared about 1.8 million years ago and was probably designed and made by *Homo erectus,* the dominant human species at that time. The tool is roughly oval, usually with one end more pointed than the other, and—this is the enigmatic part—it is carefully sharpened around its entire perimeter. Although called a hand axe, it would actually lacerate any hand that used it to carve, slice, or pound. How might these peculiar implements be linked to ice age climate? The progressively more arid climate in Africa, especially during glacial periods, would have made lakes and watering holes magnets for large numbers of game animals—a concentrated food source for early humans. But it would not have been easy to approach these herds or single out individuals for attack. In East Africa, thousands of Acheulean hand axes have been found littering the ground around the shores of lakes and in ancient, now-dry lake beds. Could they have been hunting weapons?

In 1979, an undergraduate student at the University of Massachusetts, Eileen O'Brien, made a fiberglass model of the Acheulean hand axe and discovered that it had such distinctive aerodynamics that, when thrown, it always oriented itself vertically part way through its flight. The experiment has been repeated and confirmed. One (or many) of these hand axes, thrown by *Homo erectus* into a herd of animals crowded together at a watering hole, would have landed, spinning vertically, and sliced into the back of one of the herd. It wouldn't even require very accurate throwing. Critics claim that it also wouldn't have caused a large animal much injury, but supporters of the idea claim that such attacks would have caused some animals to stumble, panicked the herd, and at least increased *Homo erectus*'s chances of securing a few meals.

So it is possible that even tools like the hand axe can be related to the ice age—the drought cycles of the fluctuating climate provided the best opportunities for their use. And through this tool, the link can be made to evolution. Once our ancestors learned that thrown objects were useful for hunting and also kept them out of harm's way, greater emphasis could be put on accuracy. Hand axes would be fine for a herd but not very useful for small groups or single animals, which would need to be precisely targeted. And the coordination required for accurate throwing went hand in hand with larger brain capacity. In turn, a bigger, high-energy-consumption brain required a reliable food supply, reinforcing and extending the importance of the newly acquired skills.

Whatever the real purpose of the hand axes, they must have been very effective, because they are found in the fossil record over a span of more than a million years, and they spread from Africa to Europe and Asia—always with the same basic shape. That is perhaps the strongest argument for their use as throwing weapons, because it is the basic form that gives hand axes their aerodynamic properties.

By the time the hand axe appears on the scene, humans had already migrated—or perhaps *spread* is a better word—out of Africa onto other continents. The routes they took were most likely influenced by the ice age climate. Europe was repeatedly being inundated by continental ice sheets at this time, its vegetation zones shifting back and forth as the glaciers advanced and retreated. There is no evidence of humans there, despite much searching. However, a fossil *Homo erectus* individual dated at 1.8 million years has been found in Georgia, at the eastern end of the Black Sea, and there are other hominid fossils in Indonesia and China of similar or slightly younger age. All the evidence suggests that hominids turned right when they left Africa, avoiding the extremes of European ice age climate and opting for the more equable regions to the east and south. Population fragmentation, and probably the emergence of new species, resulted from this dispersal, but none of the new species survived. Our own species, *Homo sapiens,* appeared much later, not in Asia or Indonesia, but in Africa, some time around 150,000 years ago. This age comes from the dating of fossils, and has recently been corroborated by DNA evidence, which can be used, in principle, to pinpoint the time when one species branched off from another.

When *Homo sapiens* arose 150,000 years ago, the Earth was close to the maximum cold of the last but one glaciation of the Pleistocene Ice Age; some 30,000 years later, by about 120,000 years ago, the ice had receded and the climate was much like today's, or perhaps slightly warmer. However, the archeological evidence indicates that our species remained in Africa through the interglacial warm interval and did not spread out until much later—probably about 70,000 years ago. By then the climate had again swung back into a another glacial period— northern Europe and North America were ice-covered, and the sea level was low, because large quantities of ocean water were frozen into the continental ice sheets. As happened during earlier waves of hominid migration, *Homo sapiens* first ventured into the Middle East, and then to India and Asia. Only later did our species enter Europe, and then from Asia, not directly from Africa. Aided by the lowered sea level—which meant that there were land bridges or only short stretches of water to cross in South East Asia and Indonesia—*Homo sapiens* had reached Australia by about 65,000 years ago, and New Guinea by 45,000 years ago.

By the time our species reached Europe, there had already been spectacular changes in human behavior and culture, even compared to direct African ancestors. Early cave paintings in France and Spain are rightly recognized as beautiful art, not crude petroglyphs. The rapid evolution continued after their arrival, and many anthropologists have argued that the pace of evolution over the past 50,000 years or so has been nothing short of phenomenal. But the details are still controversial, because in physical terms, including the size of their brains, humans were "modern" much earlier. Sophisticated toolmaking, often entailing assembly from multiple components and probably requiring some degree of complex thinking, had emerged long before, nearly 150,000 years before *Homo sapiens* appeared on the scene. Unfortunately, culture

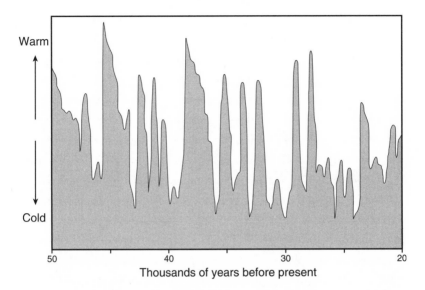

Figure 24.   There seem to be climate cycles at almost every scale that can be examined. This graph shows temperature fluctuations as recorded in a central Greenland ice core between fifty and twenty thousand years ago, a time of rapid evolution of *Homo sapiens.*

and behavior and language are not fossilized; they have to be inferred from scattered artifacts. Many workers have concluded, however, that the complex language and thought that characterize humans today have developed only over the past 50,000 years or so. William Calvin and others believe that no additional increase in brain size was required, because the necessary mental activities simply took over parts of the brain that were already there, developed earlier for other functions, such as accurate throwing. And while not everyone agrees, a case can be made that the abrupt, whiplash temperature fluctuations of the past 50,000 years of the Pleistocene Ice Age, dramatically revealed in the Greenland and Antarctic ice cores, played a part in forcing this evolution (figure 24).

It would be foolish, of course, to try to squeeze every aspect of human evolution into a theory that singled out the Pleistocene Ice Age as the only cause. There have been multiple influences at work, as well as a strong element of chance. We know that by the height of the most recent glacial advance some 20,000 years ago, *Homo sapiens* had already spread to most continents, and our species had learned to live in the harsh climate of Europe, to make clothes and to build shelters against the cold. But in spite of our dispersal around the globe, the human population remained small until quite recently. In such an environment, a series of abrupt climate changes of the kind exemplified by the Younger Dryas must have led to repeated fragmentation and sometimes demise of individual populations. In the end, though, *Homo sapiens* emerged from this trial-by-climate with enough innovations that, eventually, climate no longer mattered in the big picture of species survival, although locally it could (and can still, even today) have devastating effects.

If ice age climate has played a significant role in human evolution, a crucial question is, Why are we, among all the closely related primates, the only ones to have developed a large brain, complex language, and all the other traits of the modern human race? Why didn't the plunge into an ice age affect the evolution of other apes similarly? At the very simplest level, it may be that bipedalism—Stanley's terrestrial imperative—was the key. Chimpanzees and others remained arboreal. Their populations probably declined as the rain forests of Africa shrank in size, but they were still able to travel through the forest canopy and find enough food there to survive. *Australopithecus,* on the other hand, relied on the trees of open woodlands for safety, and as dryness increasingly fragmented these woodlands, their populations became fragmented too, and they eventually died out, giving rise to the fully land-living *Homo erectus.* The terrestrial environment had potential advantages but also dangers for this new genus. According to Calvin, Stanley, and some others, surviving in this new ecological niche required new skills, necessitating more neurons in the brain, conditions that shaped human evolution. And the continuing rapid variations of ice age climate cycles forced and sped up the process, leading in a very short time to *Homo sapiens.*

Although for us human evolution is perhaps the most interesting example of the effects of ice ages on evolution, it is far from the only one. Recent research, especially research using genetic techniques, shows that many plant and animal populations that lived during the Pleistocene Ice Age were greatly affected. Particularly at the higher latitudes of Europe and North America, the present-day distribution of many species is the result of normal biological diversity strongly influenced by climate changes and geography. DNA analysis has been especially valuable for working this out, because it allows genetic diversity within a species to be examined and specific lineages to be tracked. As might be expected, the ranges of individual species expanded and contracted during the warm-cold cycles of the Pleistocene Ice Age, but an interesting discovery is that the expansion phases were particularly important for establishing the present distributions. As the climate quickly warmed during the "rapid terminations" that characterize the end of glacial periods, small groups of "colonizers" from a shrunken original population would establish themselves and quickly expand into areas of newly suitable habitat. Sometimes they would take over very large areas with little or no competition. Later "invaders" from the original population had little chance of making a significant imprint on the genetic makeup of the exponentially expanding population of colonizers. Repeated rapid climatic cycles accompanied by these kinds of population changes led to significant genetic differences within species, and sometimes to entirely new species. Such histories are now well documented for many plants, animals, and insects, and they are relevant to our discussion of human evolution, because they are very much akin to the boom-and-bust scenario championed by William Calvin.

A concrete example comes from studies of the common grasshopper. Today, in terms of their genetic makeup, there are about five different populations of this insect in Europe. But just one of these dominates a very large area, all of northern Europe and south into the Balkans. During the last glacial maximum, 20,000 years ago, the vegetation on which grasshoppers feed was restricted to a few regions of southern

Europe in the Balkans, Spain, and Italy (this can be deduced from analysis of pollen in lake sediments). The implication of the current population distribution is that Balkan grasshoppers were the colonizers as the ice retreated and suitable habitat opened up rapidly to the north. They expanded very quickly into this new ecological niche, with no significant barriers to their mushrooming population. Groups of grasshoppers that had been restricted to Spain and Italy during glaciation, however, were limited by geography as the climate warmed—the Pyrenees and the ice-capped Alps, respectively, prevented them from dispersing northward as rapidly as their Balkan cousins. The genetic makeup of many other species in Europe and North America, from hedgehogs to fish to bears, also shows distinct geographic groupings that can be linked directly to ice age climate variability. Both in simulations and in real life, the colonizing populations that expand rapidly to fill new environmental niches have a relatively homogeneous genetic architecture, while the parent population retains greater diversity. This is paralleled in modern humans by the observation that there is greater genetic diversity among the African population than is present, for example, in modern Europeans.

It seems clear that the Pleistocene Ice Age has played an important role in the evolution of humans and other life of Earth. Before closing this chapter, it is worthwhile to ask the question, Did earlier ice ages also influence evolution? It is tempting to assume that they *must* have had evolutionary consequences—after all, some, like the Snowball Earth episodes, were much more severe than the current ice age. But the great difficulty in investigating this question is the sparseness of the fossil record. While ever more sophisticated analytical techniques allow us to document environmental changes in the distant past, it is difficult to tie the comings and goings of an individual species or genus, or even a family, with specific changes in climate. There is nothing remotely resembling the detailed, continuous ice-core and deep-sea sediment records that are available for the entire duration of the Pleistocene Ice Age. Instead, we have to be content to examine sedimentary rock sections from scattered locations around the world, a few hundred meters here, another few hundred there, and patch them together into a coherent and, one hopes, more or less continuous record. The problem of cause and effect becomes much more acute than it is for the Pleistocene Ice Age. And while there is a suggestion that the rate of extinction— usually measured in the fossil record as the fraction of existing species that became extinct during a given time period—increased during the Late Paleozoic Ice Age that occurred around 300 million years ago, the evidence for the earlier ice ages is even more nebulous.

However, some of those working on the Snowball Earth glacial episodes point to changes in the fossil record that followed these intervals as an indication that they did indeed influence the course of life on Earth. The problem is that the Snowball Earth ice ages occurred over a long period of time, between about 800 and 600 million years ago, and the appearance of new life forms in the fossil record occurred millions of years later. Whether or not they can be linked is still an open question.

Before the Snowball Earth ice ages began, the only living things that inhabited our planet, as far as we know, were one-celled organisms— bacteria and algae—and the slightly more advanced eukaryotes, which had cellular nuclei and more complex internal structures. The fact that eukaryotes survived the Snowball Earth interval, with the oceans frozen from pole to pole, has been touted by critics of the idea as evidence that no such episode occurred. How could life survive under such condition? This might be a powerful argument if every square centimeter of the planet had been frozen solid. However, even the proponents of a "completely frozen" ocean during Snowball Earth point out that there would have been active volcanic regions analogous to Iceland or Hawaii that might have sporadically or even continuously provided small regions of warmer, open water where eukaryotes could endure. In addition, open cracks and leads in the sea ice would likely have been present in the tropics, as they are in the present day Arctic Ocean in summer. Even in the coldest winter months there are fairly large patches of open water that occur regularly in the arctic, always in the same localities, presumably due to winds and currents. They are called polynyas, and they attract an abundance of bird and animal life. Thus it doesn't seem too difficult to imagine eukaryotes (and bacteria and algae too) surviving even severe global glaciation. But, like *Australopithecus* or the hedgehogs or grasshoppers of the current ice age, their habitats would have been restricted. They would have retreated to *refugia*, the term that paleontologists and ecologists use to describe places that provide islands of habitable environment during times of crisis. The sparse fossil evidence does seem to corroborate this view; it shows a decrease in diversity among the eukaryotes at about the time of the Snowball Earth glaciations. The climatic shocks of extreme cold during these periods, possibly followed by extreme warmth during short "super greenhouse" intervals, would presumably have led to the same kinds of expansion, contraction, and speciation among the eukaryotes as is seen in plants and animals during the current ice age.

Shortly after the end of the Late Proterozoic glacial episodes, a completely new type of fossil appears in sedimentary rocks, and it is these that have caused most speculation about the effects of Snowball Earth on evolution. The new organisms are called the *Ediacarans,* after the locality in Australia where they were first identified. They were soft-bodied, large (at least compared to the single-celled creatures that had dominated the oceans for billions of years before them), and displayed a range of body shapes. They appear quite abruptly in the fossil record, diversified rapidly, and then disappeared almost as quickly. Not long afterward, about 540 million years ago, the famous "Cambrian explosion" occurred, documented by the sudden appearance in sedimentary rocks of a wide range of creatures that had begun to make shells, carapaces, and skeletons—and are thus much better preserved as fossils than their soft-bodied predecessors.

The connection between the appearance of these new organisms and the Late Proterozoic Snowball Earth episodes is tenuous, but because ice ages are rare in the Earth's history and the bursts of evolution followed a series of these rare events, it is just possible that the cold intervals played a part in shaping the evolutionary pathways. But much more evidence will be required before that link can be made with any certainty.

CHAPTER ELEVEN

The Last Millennium

We are now about twenty thousand years past the peak of the most recent glacial advance of the Pleistocene Ice Age and are in the midst of the maximum warmth of an interglacial period. Record high temperatures have been the norm in recent years; newspapers report the melting of permafrost in Alaska and the possibility of an ice-free Arctic Ocean at the North Pole. And yet in the winter of 2000–2001, Scots were delighted when a spell of very cold weather allowed them to hold curling tournaments on lakes that had not been frozen for decades. In 2003, there was snow in New York in early April, a rare event by historical standards. Several years ago, a cartoon in a U.S. newspaper poked fun at the concern over global warming. It showed a bundled-up man with mittens and a scarf reading a sign, posted on a closed door, that read "Lecture on global warming cancelled due to cold weather." The same cartoon could easily have been recycled during the arctic cold spell that gripped the northeastern United States early in 2004. These things remind us that there is no such thing as average weather. Weather is what we experience daily, and it can be misleading, because we are impressed most by the extremes. Only over a longer timescale, by observing trends in temperature, precipitation, and windiness, can we really determine the direction in which the climate is moving. It is the ability to provide us with a long-term record of climate that makes cores from glaciers, lake and ocean sediments, and even trees so very valuable. We have already seen that they portray a Pleistocene Ice Age consisting of a seemingly endless series of cycles, cold succeeding warmth and dry periods succeeding wetter ones. Change, often very rapid change, is the dominant theme. It seems to be most frequent and most severe during transitions between glacial and interglacial periods, but it is omnipresent. Quite often the change is regular, following the astronomical cycles, but sometimes—especially at shorter timescales—it is chaotic. Given our uncertain understanding of the causes of climate change, this makes short-term predictions of the future somewhat hazardous.

We do not know if frequent shifts in climate also characterized earlier ice ages, or if the long, warm intervals between the Earth's major ice ages were monotonously stable, because we don't have any high-resolution records from those times. However, we do have a very high quality record of climate through the last few millennia of the Pleistocene Ice Age, and we also have written and archeological evidence from the same time period. It is very instructive to compare the two, for it provides insight into the intersection between the continuously varying ice age climate and developing human civilization.

There are now several good examples of past civilizations collapsing as a result of abrupt climate change. In most cases, the archeological evidence for collapse has long been known, but the causes have not. Many different factors—competition, war, technology development, disease, and others—have been proposed as being important, but none have been entirely convincing. The evidence of a climate change connection has come from improved techniques for measuring or inferring local environmental conditions such as precipitation and temperature, and from new information about the timing of changes in these conditions. And although temperature is often the parameter that attracts the most attention in climate records, in many cases, rapid swings in the amount of precipitation have been most important. A classic example comes from the Akkadian civilization of the Middle East, which was thriving before it experienced sudden collapse about 4,200 years ago. More than a century of archeological research had documented this decline in great detail, but there was no agreement about its cause. However, on the basis of regional climate data that are now available, it appears that the collapse can be linked to an abrupt transition into a time of severe drought. The Akkadians had adapted to life in an arid region, but they still depended on predictable seasonal rainfall for irrigation. When the regional climate changed and the rain suddenly failed, they were in trouble. Whole cities were abandoned and a prosperous empire crumbled. Hordes of refugees moved southward, in what is present-day Iraq, from northern Mesopotamia to regions along the Euphrates River that had a more dependable water supply. But the drought was widespread, and the "barbarians" from the north—as they are termed in the writings found on clay tablets from the region—so burdened already stressed resources that these cities too were brought to the verge of collapse. A very similar fate appears to have befallen the Mayan civilization of Central America several millennia later. Like the Akkadians, their society collapsed rapidly for reasons that are hard to discern from archeological evidence alone. But they too depended on seasonal rainfall for irrigation and survival. Sediment cores from the nearby Caribbean show that between 1,100 and 1,200 years ago, the region experienced an extended period of drought, punctuated by short but repeated spells of even more severe aridity. That is precisely the time when Mayan civilization collapsed. Like the Akkadians, they were unable to sustain their formerly prosperous cities when the rains failed.

The Akkadians and the Mayans present interesting but isolated examples of the influence of climate on mankind. The periods of drought that affected them may have been local, but they were most probably related to more global shifts in average climate as the Earth emerged from the most recent glacial period. However, what has really caught the interest of a small but growing group of anthropologists, archeologists, and historians who are examining history from the perspective of climate is the past 1,000 years. This period has the advantage that there is an abundance of proxy records from ice cores, tree rings, and sediments, for the most part providing very high time resolution. In many cases, even year-to-year changes can be tracked with confidence. Furthermore, climate information can be compared directly with the written historical record. Temperatures deduced from central Greenland ice cores can be checked against statistics for wheat harvests in Europe or the northern limits of vineyards. Perhaps not surprisingly, many intriguing coincidences of timing have been discovered between climate change and historical events.

In the climate record of the past millennium, two periods stand out— a warm one that has been called the "Medieval Warm Period," and a colder interval dubbed the "Little Ice Age." The latter is a misnomer in technical terms, because it was far from being an ice age. In reality, it was nothing more than a slightly cool interval within the current inter-glacial period, but it was characterized by expansion of glaciers in many parts of the world—this was especially obvious in the relatively heavily settled Alps—which led to the name. Compared with the longer-term fluctuations of the 100,000-year ice age cycles, or even shorter-term shifts such as those of the Younger Dryas period, these two slightly abnormal intervals experienced only small changes in average temperature, little more than one or two degrees Celsius. However, the historical record indicates that even fluctuations of this magnitude had direct, and some-times severe, consequences for human society. Again, it was not just the temperature change that was important, but the overall weather patterns that accompanied the change—especially the amount and distribution of precipitation, and the frequency of storms.

Eight or nine hundred years ago, Europe was in the midst of the Medieval Warm Period. Although there were the inevitable spells of "bad weather," summers were generally warm and crops plentiful enough that there were few serious famines. Temperature data from Greenland ice cores corroborate the inference made from the written record that this was a time of relatively warm, benign climate in the region. Vineyards flourished in England to the point that the French became concerned about the impact on their own wine industry. In the North Atlantic, ice packs retreated northward, especially in summer, and the incidence of severe storms decreased. The Vikings, who were already expert sailors and had raided coastal Ireland, Britain, and Europe centuries before, sailed north and west under the more favorable conditions and established settlements in Iceland and coastal Greenland. They conducted regular supply and trade voyages between these colonies and Scandinavia, and from Greenland they also explored the eastern shore of North America. Although they never colonized the New World, they did found small settlements about 1,000 years ago, such as the one that tourists now visit at L'Anse aux Meadows, in northern Newfoundland. In Europe, the warm and equable climate meant that kings and landowners prospered, and laborers and peasants rarely went hungry. The amount of land devoted to agriculture increased, and, especially in northern countries, fields crept up hillsides to elevations that had not been farmed before. The range of warmth-loving crops moved steadily northward. The general prosperity meant that there were funds to fight the Crusades and to employ generations of stonemasons who traveled throughout Europe building beautiful, massive, and very expensive cathedrals. In harsher times, such feats might not have been possible.

Europeans living at this time could not have anticipated the rigors that the Little Ice Age was about to bring. As the thirteenth century drew to a close, the Medieval Warm Period was also nearing its end. The Greenland ice cores, together with other climate records, show that temperatures decreased beginning around 1300 and stayed low—although with some intervening warm intervals—well into the nineteenth century (figure 25). Where, exactly, to place the beginning and end of the Little Ice Age is somewhat subjective. Some scientists restrict the term to the period between 1600 and 1800, when Europe experienced some especially cold weather and Alpine glaciers grew substantially (in the early 1600s, one observer reported from Switzerland that a glacier was advancing *daily* by as far as one could shoot a musket). Other climatologists take the 1600–1800 period simply to be the climax of the Little Ice

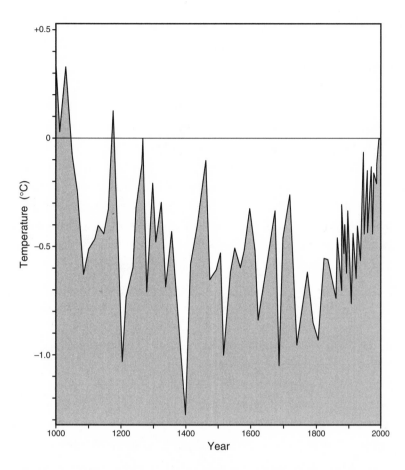

Figure 25. Temperature variations through the last millennium as recorded in a Greenland ice core. The scale is relative, with today's average fixed at zero. The Little Ice Age lasted from approximately 1300 to 1850, and its coldest period was near its end.

Age, and use the term for the entire time from 1300 until about 1850, when average temperatures began a slow climb toward today's values. But however one defines the Little Ice Age, throughout even this geologically short timespan, there were multiple temperature excursions bringing both warmer and colder weather. It was only in an average sense that the interval was cold.

What caused the Little Ice Age, or even what caused the short-term cold-to-warm-and-back-again flip-flops within the Little Ice Age, is unknown. Climate science is notoriously difficult, because there are so many interconnected variables at work that cause and effect are often impossible to discern with confidence. The oceans, however, are thought to play a central role, because they store a vast amount of heat energy, which they disperse worldwide through the mechanism of ocean currents. The atmosphere is important too, and there are close connections between what happens in the oceans and the atmosphere. In recent years, continuous global monitoring of winds, ocean temperatures, clouds, and many other parameters that affect weather have provided a good empirical view of how some aspects of the climate system work. However, a clear understanding of the causes is still elusive. A good example is the El Niño phenomenon—over the past decade, a very detailed picture has emerged of how ocean temperatures and atmospheric pressure change in the equatorial Pacific Ocean during El Niño cycles, yet exactly why this happens, or why the cycles occur every three to seven years, or how events in the equatorial Pacific affect the weather in the United States or Australia or India is not well understood. In the North Atlantic region, where most of the historical records of the Little Ice Age are found, there are also periodic shifts in the overall weather patterns. They seem to occur somewhat less regularly than the El Niño events of the Pacific (at least over the relatively short span of time they have been observed), and they last longer. These climate changes are related to a phenomenon known as the North Atlantic Oscillation—NAO—and they bring Europe, and eastern parts of North America, alternating periods of warmer and colder average temperatures. They also have a significant effect on the distribution of rainfall. The "oscillation" part of the NAO refers to variations in the strength and position of atmospheric high and low pressure zones in the North Atlantic, which in turn affect the location of the jet stream and the path that winter storms follow across eastern North America, the North Atlantic, and Europe. Often one mode of the oscillation dominates for a decade or two, followed by an abrupt shift to the other. Although the NAO is defined in terms of an atmospheric phenomenon, many researchers contend that it is closely linked to changes in the circulation pattern of cold and warm waters in the North Atlantic Ocean, again emphasizing the close coupling between the ocean and atmosphere. Most likely prolonged times of warm, stable climate such as the Medieval Warm Period, or generally cool and wet periods such as the Little Ice Age, are manifestations of something like the different climate modes that characterize the present-day North Atlantic Oscillation.

Even if the prevailing weather of the Little Ice Age can be linked to an observed phenomenon such as the North Atlantic Oscillation, however, that still doesn't solve the problem of cause. For that it's useful to recall James Croll's maxim, which was that in order to understand nature, one has to understand the underlying physical laws. Croll realized that ice ages might occur when lower than normal amounts of solar energy are received on the Earth because of orbital changes. Heat energy from the sun is also what drives atmospheric winds and ocean currents and is ultimately responsible for phenomena such as the North Atlantic Oscillation. Recognizing this, scientists have been searching for clues about how the input of solar energy might have varied over the timescales of the Medieval Warm Period and the Little Ice Age. Orbital parameters of the kinds investigated by Croll and Milankovitch occur much too slowly to explain these shorter-term climate fluctuations, so evidence for other types of variation has been examined.

Climatologists are nothing if not ingenious. They have developed a bevy of proxies to investigate past climate, and several of them provide good measures of the sun's energy output. Although it doesn't fluctuate very much, the sun's "activity," a measure of how much heat energy it releases into space, does vary. A well-known example is the eleven-year solar cycle, the peak of which is marked by intense solar storms, great eruptions of matter in flares from the sun's surface that hurl ions out into space—a symptom of a more energetic sun. The magnetic fields that accompany the solar flares play havoc with telecommunications on

Earth, and those who live at high latitudes get to see spectacular displays of the aurora borealis. Times of high solar activity are also marked by sunspots—phenomena that have been observed and recorded for centuries, initially because they were thought to be portents of the future, and later out of scientific interest. Two thousand years of combined and (almost) continuous Chinese and European observations paint a vivid picture of the sun's activity. They show that the number of sunspots recorded during the Little Ice Age was low, by implication the sun ever so slightly cooler. The change is small enough that it cannot, on its own, account for the climate variations. But even variations of this magnitude may be part of the explanation, through what climatologists refer to as forcing. Just as orbital changes do on a longer timescale, they seem somehow to pace the temperature changes associated with events such as the Medieval Warm Period or the Little Ice Age.

Other proxies confirm the sunspot observations. When the sun's activity is high, the strength of its magnetic field is also high, and cosmic rays (high-energy particles that originate outside the solar system and permeate interstellar space) are deflected away from the Earth in greater numbers than usual. This happens because cosmic ray particles carry an electric charge, and when electrically charged particles enter a magnetic field, their direction of travel is changed—they curve away from their original path. Normally, a combination of the sun's and the Earth's magnetic fields deflect away most cosmic rays, but some especially energetic ones are only minimally diverted and penetrate through to the atmosphere. When this happens, they smash into nitrogen and oxygen atoms and make radioactive isotopes such as carbon-14. The amount of carbon-14 formed is thus a fairly sensitive measure of solar activity—the stronger than normal magnetic field that characterizes times of high solar activity deflects more cosmic ray particles away from the Earth, resulting in lower carbon-14 production. Tracing how carbon-14 has changed in the atmosphere, which can be done by measuring the isotope in ice cores or tree rings, is therefore another way to gauge the sun's activity. The carbon-14 data corroborate the sunspot record and indicate that solar activity was generally high during the Medieval Warm Period, and low during the Little Ice Age.

As solar activity decreased near the beginning of the Little Ice Age, the signs of a deteriorating climate in the North Atlantic region are abundantly clear both from scattered historical records and archeological evidence. In the islands of the Canadian arctic, the Inuit people began to migrate south. By the 1340s, sailors traveling between Iceland and Greenland had to follow longer, more southerly routes in order to avoid ice and treacherous weather. The Norse settlements in Greenland went into decline and were eventually abandoned as an already unforgiving environment became even harsher. Skeletons from graves at some of the Greenland settlements show that by 1500, near the end of the settlement period, even the average size of the few people still living there had decreased significantly. In the worsening weather, farming became increasingly difficult, and their diets correspondingly less varied and nutritious. In the low-lying coastal areas of Germany, Denmark, and Holland, fierce storms obliterated large tracts of agricultural land in the thirteenth and fourteenth centuries, and by most estimates claimed more than a hundred thousand lives. And in the Alps, glaciers began to encroach on farms and pastures that had been established during earlier, warmer times.

In spite of these undeniable facts, there is still much debate about the degree to which historical events such as economic decline, famine, the "Black Death" (plague), and changes in political fortune were due to the deteriorating climate as opposed to other factors. Perhaps climate is just a "forcing factor," in the same way that the sun's activity or the Earth's orbital parameters are forcing factors for climate itself, not wholly responsible, but important enough to tip the balance in one direction or another. However, some historical events of the Little Ice Age are so widespread throughout Europe that an external factor such as climate seems likely to be the predominant cause.

An example is the disaster that struck across Europe during the early years of the fourteenth century. The relative prosperity and benign weather of earlier times had resulted in an expanding population. Much marginal land had been occupied and farmed. Then a series of cold, wet years struck. Grain rotted in the fields, or sometimes couldn't be planted at all. The cost of food rose rapidly. Livestock and people began to succumb to famine and disease, and even the rich had difficulty getting enough to eat. Social unrest was rampant; gangs of desperately hungry peasants roamed the countryside searching for food, and in some places there were reports of cannibalism. By the 1320s, land and even whole villages had been abandoned throughout Europe, and population fell precipitously, in places by more than half. This was indeed abrupt change; it occurred within a human lifetime, at a time when life expectancy was half what it is today. A few decades later, the first of a series of waves of bubonic plague, the infamous Black Death, hit an already reeling Europe. Huge segments of the population, especially in crowded urban areas, succumbed to the disease. The first half of the fourteenth century is also the period that historians identify with the dismantling of the feudal system that had prevailed throughout Europe. Can all of this be ascribed to climate? Certainly, the poor harvests of the early 1300s, and the fungus and other diseases that affected crops, can be attributed directly to a series of very wet and generally cold years. People and livestock, weakened by shortages of food, were especially susceptible to disease. The Little Ice Age was in its early stages, but already, it seems, it was having a profound effect on society.

The impact of the changing climate was not just felt on land either. At the time of the Little Ice Age, cod was an important source of protein throughout the North Atlantic region, especially for coastal populations. Abundant, large, and nutritious, it was in great demand, in part because the Catholic Church allowed its adherents to eat fish, but not red meat, on Fridays. It was also a permitted food throughout Lent. It was easily preserved; salted cod is light, has a high food value, and can be kept for long periods without refrigeration. Long before the Little Ice Age, Norse explorers had carried salt cod as their principal food on sea voyages. It became an essential and very valuable commodity, and as a result the cod fishing industry was large and extremely competitive. As the Little Ice Age began, fish seemed to be there for the taking, not subject to the same season-to-season vagaries of the weather that affected crops and livestock. But that optimistic outlook was soon to change.

Cod have a limited tolerance for temperatures outside a narrow range of about four to seven degrees Celsius. Below two degrees they suffer kidney failure. Because of this, their geographical range, especially at the cold northern extremes, is particularly sensitive to changes in water temperature. Unfortunately, existing records are not detailed enough to trace exactly how cod populations shifted during the Little Ice Age. However, several general trends are clear. As the northernmost Atlantic progressively became ice-filled due to the colder weather, the cod moved south. Catches around Greenland decreased, as did those near Iceland. Fishing off the coast of Norway also became more difficult. The fish didn't completely disappear from any of these localities, but their numbers dwindled, and it became harder and harder for the Basque, English, and Dutch fishing boats that regularly visited these waters to fill their holds with cod. As a result, and in spite of the generally stormier weather, they began to search farther afield. When their explorations first led them to the coast of North America is not known—then as now, fishermen didn't want to give away their secrets. But in 1497, the Italian explorer Giovanni Caboto, known to the English as John Cabot, sailed along the eastern coast of Canada and reported waters teeming with cod. In places, they were abundant enough to literally scoop out of the sea with baskets. Cabot noted that the sea was also full of Basque fishing boats. This was only a few years after Christopher Columbus "discovered" North America; the fishermen had certainly been there long before. Just as the warmth of the Medieval Warm Period had allowed the Vikings to sail westward to North America, so the cod populations, responding to the cold of the Little Ice Age, led Europeans across the Atlantic to the New World.

Although the Little Ice Age was, on average, significantly colder than the immediately preceding and following periods, it was by no means a single, uninterrupted interval of cold. There were reprieves, sometimes fairly long ones. There seem to have been significant spikes upward in temperature in the early 1500s and again in the early 1700s, each followed by a return to much colder times. There are hints that cod migrations, political crises, and the frequency of famines were all affected by these fluctuations. In an especially cold period through the mid and late 1600s, cod essentially disappeared from the water around the Faeroe Islands, where they had previously been abundant. About the same time, residents of the Orkney Islands off the north coast of Scotland were several times startled by the improbable appearance on their shores of Inuit people in kayaks, presumably forced south and east by enlarging ice packs to the north. Poor harvests and harsh weather, together with the promise of better life elsewhere, initiated emigration from Scotland that would continue for centuries. Perhaps one of the greatest difficulties faced by people of the time was unpredictability. While it might be possible to adapt agriculture and housing to consistently lower temperatures, it was very difficult to cope with wild swings between very cold and very warm. And weather records, which are quite complete for much of Europe and parts of North America from the early 1600s on, show that climate extremes were often juxtaposed. The coldest winter in central England between 1659 and 1979 was the winter of 1683–84; one of the warmest was just two years later in 1685–86.

Not all of the news from the Little Ice Age was bad. Countries such as Holland benefited from the shift of fish stocks southward into the North Sea. In response to coastal flooding, exacerbated by the violent storms of the sixteenth and seventeenth centuries, the Dutch also became experts in the technology of reclaiming low-lying land from the sea. Subsistence farming, always risky at the best of times, declined during the Little Ice Age, and more specialized agriculture sprang up, especially in Britain. Growing cash crops for the city was usually a better way to survive than trying to be self-sufficient. Even glass windows are believed by some to be a response to the cold temperatures of the

Little Ice Age—they helped keep the cold out but still provided a view of the outside world. And while no one has yet suggested any connection between climate and the American Revolution, it is quite likely that the chronic shortages of bread and grain in France in the late 1700s, which were due at least partly to the bitter winters, unpredictable climate shifts, and generally bad weather of the time, fed the underlying discontent that led to the French Revolution.

The Little Ice Age even influenced art. Snowy winter scenes suddenly appear abundantly in sixteenth- and seventeenth-century European paintings. Skaters glide on canals and lakes that have not been frozen in living memory (figure 26). In 1970, Hans Neuberger, a meteorologist at Pennsylvania State University, analyzed more than twelve thousand paintings from American and European museums, all dating between 1400 and 1967. He classified them by region and date, and looked at them with a meteorologist's eye. Neuberger was particularly interested in depictions of the sky, and while not every painting was amenable to this kind of examination, many were—even if they only showed a glimpse of the sky from a window. While artists surely take some license with their subjects, Neuberger's analysis uncovered some clearly defined trends—for example, none of the British paintings he studied showed a completely clear sky, while some 12 percent of those from Mediterranean countries did. Half of the British paintings showed the sky completely overcast, a higher fraction than any of the other regions he studied. His data show an increase in cloudiness between 1400 and 1550, and then an abrupt further increase—more than 50 percent—especially in the abundance of low, scudding clouds. Cloudiness peaks during the seventeenth century but remains high throughout the Little Ice Age. Neuberger also pointed out that most paintings from the last few hundred years of the Little Ice Age—on average its coldest period—are very dark compared to both earlier and later art. This could just be a popular stylistic device, but Neuberger suggests that it might also be a reflection of the generally cloudy, low-illumination conditions that prevailed.

Figure 26.   Painting by Sir Henry Raeburn of the Reverend Robert Walker skating on Duddingston Loch, near Edinburgh, Scotland, toward the end of the Little Ice Age. Duddingston Loch rarely freezes in winter now. Reproduced with permission of the National Gallery of Scotland.

And in 2003, Henri Grissino-Mayer, a tree-ring expert at the University of Tennessee, and Lloyd Burckle, at Columbia University, suggested yet another possible connection between the arts and the Little Ice Age: they proposed that climate may be partly responsible for the exquisite sound of Stradivarius violins. By examining tree rings,

Grissino-Mayer found that growth in European high-altitude forests slowed because of the cold of the Little Ice Age. He also discovered that between 1625 and 1720, the trees showed exceptionally narrow growth rings, producing dense and strong wood—properties that may have enhanced the quality of the instruments made by the renowned Italian craftsman. Stradivarius produced his most famous violins between 1700 and 1720.

A hallmark of ice age climate change, at least when viewed from the perspective of its impact on human societies, is abruptness. With little or no warning, there have been drastic shifts in temperature, storminess, and precipitation, both regionally and globally. Frequently, these shifts, although very rapid, leave the climate system in a new mode that persists for a relatively long time. The Mediaeval Warm Period and the Little Ice Age, as well as the occasional abrupt fluctuations that occurred within them, are good examples from the past millennium. As already discussed, there are even more striking events, such as the Younger Dryas interval, on a longer timescale. There is little doubt that similar changes will occur in the future, and understanding the underlying causes of such events is important if there is to be any hope of predicting them or mitigating their impact on society. Abrupt climate change is currently a hot topic in the environmental sciences, and a large cadre of scientists from diverse disciplines are working on the problem. In a relatively short time, much has already been learned, and although definitive answers—always elusive in science in any case—are not available, some general conclusions are.

Most of the attempts to understand why rapid climate shifts happen involve several concepts that are quite familiar. The first is the idea of a threshold, a state that, if crossed, more or less automatically shifts the climate into a different mode. To cross a threshold, however, requires something else—usually an external "forcing" or trigger, some process or phenomenon that will change the system, either slowly or rapidly, until it reaches and crosses the threshold and flips into the new mode. Such changes also usually require some type of positive feedback, a multiplying effect that ensures that the change will be global or universal. None of these ideas is particularly new—recall James Croll's belief that variations in the Earth's orbit would act as an external forcing, cooling the Earth until snow persisted on the ground throughout the year. Year-round snow was the threshold, and it would also provide positive feedback: it would amplify the cooling by reflecting more solar energy back into space, initiating rapid expansion of ice sheets and a new glacial interval. By building such ideas into complex computer simulations of the global climate—an enormous task that requires great computing power—and by using accurately determined ocean and atmospheric conditions, it has been possible to examine the effects of different types of external forcing on the entire system—a kind of "what if" approach.

One conclusion of the simulation studies, already known in a general way from earlier work, is that the ocean current system is very important for distributing heat. In particular, changes in the way ocean circulation occurs in the North Atlantic Ocean have been implicated in some of the large and abrupt temperature changes observed in the Greenland ice-core data over the past few tens of thousands of years. As discussed briefly in chapter 6, the warm surface water of the Gulf Stream moves northward in the Atlantic, evaporating and cooling as it goes. Both evaporation, which increases the salt content of the water left behind, and cooling, which contracts it, cause the density of the water to increase. By the time it reaches high latitudes, it has become so dense that it sinks, displacing the underlying lighter water. There are a few other places, such as the Antarctic, where very cold surface water also becomes dense enough to sink, but the process is most important in the North Atlantic—so important that it is a driving force in the circulation of the entire ocean. The cold sinking water spreads southward across the equator in a deep layer, south to the Antarctic and around into the Indian and Pacific Oceans. In places, it upwells again to the surface, and surface currents make up the return flow, eventually again joining the Gulf Stream to complete the circuit. If something were to shut down the sinking of North Atlantic seawater, the whole ocean circulation system would slow down and either stop completely or reorganize. The Gulf Stream would no longer carry warm tropical water into the North Atlantic. Greenland and Europe would lose the warming benefit of this current, and their climates would abruptly become much colder.

That sounds very nice and simple in theory. Could it actually happen? Many researchers now believe that just such a scenario was responsible for the Younger Dryas cold period discussed in the previous chapter, and probably also for many other cold snaps that can be identified in the Greenland ice cores. Hints can be found in sediment cores from the North Atlantic that during these intervals, the flow of Gulf Stream water slowed, and the amount of new dense bottom water being produced declined. In addition, ice cores from the Antarctic show a slight warming at high southern latitudes, an effect that has been linked with a weak or nonexistent Gulf Stream. Under conditions similar to those at present, the Gulf Stream cools the Antarctic slightly by drawing warm water out of the Southern Hemisphere and transporting it northward; if the Gulf Stream slowed or stopped, a small amount of warming would be expected.

That still leaves the question of cause. The search for reasons for the on-again off-again nature of the Gulf Stream has focused on processes that could change seawater density, because, as explained above, density plays an important role in ocean circulation. For a given batch of ocean water, density depends on temperature and dissolved salt content (and for this reason the circulation is referred to as thermohaline circulation). If some process were to decrease the density of North Atlantic surface water, it would eventually cross a threshold value and float rather than sink, shutting down the thermohaline circulation. This could happen by addition of low-density fresh water from rapidly melting glaciers, as was discussed in chapter 6. Just such a scenario has been proposed for abrupt temperature decreases recorded in Greenland ice cores near 12,800 (beginning of the Younger Dryas) and 8,200 years ago, both of which correspond to sudden changes in glacial Lake Agassiz's drainage that added large volumes of fresh water to the North

Atlantic. It is also possible—paradoxically—that present-day global warming will lead to cooler temperatures in northern Europe through a similar effect. Both accelerated addition of fresh water due to melting of the Greenland ice sheet and the general warming of seawater because of globally higher temperatures will decrease the density of surface water in the North Atlantic. There is some evidence that the amount of sinking cold water in the North Atlantic has decreased slightly in recent years, but the measurements have not been carried out over a long enough period to determine whether this is a long-term trend or just a minor deviation from the average.

Although a strong case can be made that large-scale changes in North Atlantic ocean circulation were responsible for at least some of the rapid temperature changes recorded in Greenland ice cores, there is no evidence that the less severe climate variations of the past millennium, disruptive as they were for European civilization, had a similar origin. As mentioned earlier in this chapter, one idea is that the sun's activity may have been the important forcing factor. Possibly that could have tipped the North Atlantic Oscillation into a mode that dominantly brought cold weather to the region. Even volcanic activity has been implicated, not as a cause of Little Ice Age cold, but as a process that occasionally and temporarily exacerbated the already-cool climate. That volcanic activity can have a measurable effect on temperatures worldwide is no longer in dispute—the volcanic dust and sulfurous gases blasted into the stratosphere during the 1991 eruption of Mt. Pinatubo in the Philippines so reduced solar energy reaching the Earth's surface that global temperatures were lowered by about half a degree Celsius for over a year. That doesn't sound like much, but it is a sizeable fraction of the average temperature reduction during the Little Ice Age. The seventeenth century saw at least five large, explosive eruptions, beginning with the most massive, in the Peruvian Andes, early in 1600. Ash from this eruption is easily identifiable in both Greenland and Antarctic ice cores. Records from Europe and North America count the following summer, in 1601, as the coldest for hundreds of years. In 1815, as the Little Ice Age was drawing to a close, there was an even larger eruption on the island of Sumbawa in Indonesia. Again, the following summer was frigid. The year 1816 became known as the "year without a summer." Snow fell in New England in June, and crops failed in Europe.

If there is a lesson to be learned from our knowledge of the past millennium's climate history, it is that surprises abound even for this very short snippet of geological time, which, when viewed from the long-term perspective of the entire Pleistocene Ice Age, enjoyed a relatively warm and stable interglacial climate. Modern societies for the most part are better equipped to deal with such surprises than were those of even a hundred years ago, but are not entirely immune. Just-in-time logistics systems and highly concentrated and specialized agriculture are as likely to be disrupted by abrupt climate change as some earlier technologies. Energy grids even now have difficulty coping with high demand during heat waves, when millions of air conditioners are operating at full capacity. Just as troubling is our inability to predict, even in a general way, what may happen to the climate system as a result of human influences. A great, unintended experiment in "climate forcing" is under way as we add more and more greenhouse gases to the atmosphere. Whether or not we shall reach one of those thresholds that seem to separate different climate modes, and what will happen if we do, is still unknown.

CHAPTER TWELVE

Ice Ages and the Future

It is worth reiterating here something that was pointed out in the first chapter of this book but may have drifted into the background since: the Earth is still in an ice age. We are in a warm period, one of the many interglacial intervals that have occurred throughout the Pleistocene Ice Age, but even so, there are significant amounts of permanent ice in polar regions. It is easy to forget that this may be just a short respite before another glacial interval begins. Or perhaps the respite may be longer than we anticipate. Man's activities may intervene and confound our ability to use the natural climate variations of the past as a tool to predict the future.

In spite of the uncertainty about the future, we do know that the past few million years of the Pleistocene Ice Age have been fairly unrepresentative of the long history of our planet. Over that four and a half billion years, only a few major ice ages have been identified, separated by hundreds of millions of years of warmer climates when no major ice sheets blanketed the continents. We have touched on some of the possible causes for these unusual episodes in previous chapters: variations in the shape of the Earth's orbit around the sun, and the tilt and wobble of its axis of rotation; the positions of the continents as they move about on the Earth's surface due to plate tectonics; and the concentration of greenhouse gases, especially carbon dioxide, in the atmosphere. It is widely agreed that *within* the Pleistocene Ice Age, it is the astronomical parameters that have regulated the repeated cycles of glacier growth and decline. And by analogy, the course of ancient ice ages may have been similarly controlled by these parameters. But it is not at all clear that any conjunction of these astronomical conditions would on its own be sufficient to tip the Earth into a cold period and initiate an ice age. Much more likely, most scientists now believe, is that a combination of factors—including at the least the astronomical parameters, greenhouse gases, and plate tectonics—must play a role. Under the right conditions, a relatively small change in one of these—perhaps a fluctuation in the Earth's orbit—may act as a trigger to initiate a cold interval. An additional important ingredient seems to be positive feedback, some process that amplifies the effect of the other factors.

Recent concern about global warming, and especially the impact of mankind's addition of greenhouse gases to the natural atmospheric inventory, has led to increased awareness of the role these gases may play in starting or ending ice ages. Data from ice cores have further stimulated interest, because they show that on almost every timescale, the concentration of greenhouse gases has tracked changes in temperature. Rapid addition of these substances to the atmosphere may have triggered some of the abrupt temperature rises recorded in the ice cores, and it has also become clear that for the future greenhouse gases will be a major factor in determining when the Earth will slip into the next glacial period of the Pleistocene Ice Age. It is just possible that the greenhouse gas content of the atmosphere is *the* critical necessary condition that controls when an ice age will begin or end.

What, exactly, are the important greenhouse gases, and how do they affect the Earth's temperature? There are quite a few, but the most important in terms of their ability to trap solar energy in the atmosphere are water vapor (H_2O), carbon dioxide (CO_2), and methane (CH_4). All three are molecules that absorb the heat energy radiated up into the atmosphere from the Earth's warm surface. The surface is heated by incoming solar energy—the energy you feel as you lie on a beach in the sun—which is short-wavelength radiation that is not absorbed by the greenhouse gases. Because of their structures, these gases only absorb energy of specific longer wavelengths, which coincidentally match those of the energy radiated away from the heated surface. The greenhouse molecules not only absorb this energy, they also re-radiate much of it back into the atmosphere. Although the analogy is not quite perfect, they act something like window glass that lets the sun's energy in but blocks the heat in a warmed-up room from making the return journey into the outside air.

All of the greenhouse gases are quite minor constituents of our atmosphere, but without them much of the solar energy incident on the Earth would be lost almost as quickly as it is gained, and we would have a much colder home—with an average temperature of about $-18°C$, far below the freezing point of water. Of the three top greenhouse gases, water vapor is by far the most important. However, the atmosphere's water vapor content tends to follow temperature changes rather than cause them, because higher temperatures promote evaporation and thus higher water vapor concentrations, while lower temperatures have just the opposite effect. Water vapor is thus more likely to be an amplifying factor than a trigger, with low concentrations helping to maintain the cold temperatures of an ice age and much higher values during warm times tending to keep average temperatures high. Methane, on the other hand, has the potential to cause large temperature changes because it is a very efficient greenhouse gas—on a weight-for-weight basis, it is much more effective than carbon dioxide. There is currently a lively debate about whether or not changes in atmospheric methane were responsible for some of the rapid temperature fluctuations recorded in the ice cores, especially abrupt warming events such as the one at the end of the Younger Dryas cold period. As we shall see, there is good evidence that large quantities of methane have been released quickly in the past through natural processes; however, methane is a relatively unstable molecule in the atmosphere and is fairly quickly (in roughly a decade for the average methane molecule) destroyed through chemical reactions with other compounds. The effect of even a large increase is therefore of limited duration, although it has been argued that even a short-lived temperature spike could push the climate over a threshold and into a new state.

That leaves carbon dioxide as the greenhouse gas most likely to be involved in the initiation—and perhaps also the end—of ice ages. There are a number of natural processes that produce or consume this gas, and thus have the potential to change its atmospheric concentration. And unlike methane, it is not rapidly destroyed by chemical reactions with other compounds and so has a long lifetime in the atmosphere. Svante Arrhenius, as we saw earlier, recognized long ago that CO_2 could be important for regulating the Earth's climate. All one needs to do to understand its effectiveness is to consider our sister planet Venus. Much like the Earth in some ways, Venus has a very different atmosphere—one composed almost entirely of CO_2. There is so much CO_2 that the atmospheric pressure is almost one hundred times that on Earth, causing a "super greenhouse" that maintains a surface temperature about twice as hot as your kitchen oven turned to its highest setting. On October 22, 1975, the Russian Venera 9 spacecraft landed on the scorching surface of Venus. It reported a temperature of 485°C and an atmospheric pressure ninety times that on Earth. Although the lander had been designed to withstand high temperatures—it has been described as having been "built like a tank"— Russian scientists were nevertheless elated when the camera on board survived the roasting temperature long enough to send back the first images of Venus's rocky surface.

The sources and sinks of atmospheric CO_2 on Earth, the processes that could change its concentration and thereby put our planet on a path toward the beginning or end of an ice age, were touched on in chapter 8 in connection with Snowball Earth. In the short term, carbon dioxide is regulated by biologic processes. During photosynthesis plants take CO_2 from the atmosphere, use the carbon to make organic material, and release oxygen into the atmosphere. When the plants die, bacteria go to work to oxidize the carbon and convert it back into CO_2, which is returned to the atmosphere. The photosynthetic cycle is rapid enough that there are actually small but easily measurable annual cycles in the atmosphere's carbon dioxide content, with more CO_2 consumed in the warm, bright summer months, less in the winter. There can also be significant lags in the return of CO_2 to the atmosphere if the organic matter produced during photosynthesis is sequestered, for example by burial, and not quickly oxidized. There is some evidence that this happened starting around 400 million years ago when the vascular plants, with thick stems capable of carrying water and nutrients to branches and leaves, first evolved. Their rapid proliferation and growth of the Earth's first forests meant that large amounts of carbon-rich organic material were produced, all ultimately from atmospheric CO_2. Much of the carbon was incorporated into soil—then also a new feature of the evolving Earth—or peat that was buried and converted into coal, instead of being oxidized and returned to the atmosphere. As a consequence the CO_2 content decreased significantly. Because this happened in the run up to the Permo-Carboniferous Ice Age, it may well have been the factor that set the stage for that cold period. As we saw in chapter 8, the time when this ice age occurred is characterized in the geologic record by repeated cycles of coal deposits in some parts of the world. When we burn that coal today we are returning some of that long-buried CO_2 to the atmosphere.

Although biological sources control the short-term cycling of CO_2, in the long term the main source of CO_2 is volcanism. When volcanic lavas erupt on the surface, they bring with them a variety of gases from the Earth's interior, carbon dioxide among them. All those holes in a piece of pumice are nothing more than gas bubbles trying to escape from the frothy, molten rock as it cools, and a part of the gas in the bubbles is CO_2. Over geologic time, it is the balance between volcanic supply and consumption by the chemical weathering of rocks—which we shall come to shortly—that has provided our planet with a relatively stable climate. Venus shows us what can happen if supply far outstrips consumption. Fortunately for us, our watery planet has a mechanism for achieving a rough balance over long time periods.

The process that keeps our atmospheric carbon dioxide in check is the chemical weathering of rocks. When rain falls through the atmosphere, it dissolves a small amount of CO_2, making carbonic acid. The higher the carbon dioxide content of the atmosphere, the more acidic the rainwater—at today's levels (the atmosphere contains slightly less than 0.04 percent CO_2) rain is quite acidic. As a result, it can actually dissolve solid rock. The process is slow, and each rainstorm dissolves only a tiny amount, but over time, the erosion becomes quite substantial. If you have any doubt about this, just visit an old graveyard. Some types of rock dissolve so quickly that tombstone inscriptions become illegible in a century or less, especially in places with heavy rainfall.

It would be quite interesting to follow a molecule of CO_2 from the atmosphere through its journey to the ocean and beyond. Our molecule would form a new compound, carbonic acid, when it dissolved in a drop of rainwater. No harm done when the raindrop splattered on a rock at the Earth's surface, the molecule of carbonic acid would get to work dissolving the rock. In that process, it would release various chemical elements—calcium for example—and itself be transformed again, into a compound referred to as a bicarbonate ion, but still incorporating the original CO_2 molecule. Both the calcium dissolved from the rock and the bicarbonate ion would be present, in a dissolved state, in the moisture that seeped away through the soil and eventually made its way into streams and the ocean. Planktonic organisms living in the surface waters of the sea use the dissolved substances in seawater to make their calcium carbonate shells, so our carbon dioxide molecule might find itself combined with a calcium atom to form a molecule of calcium carbonate in one of these shells. When the organism died, the shell would fall to the seafloor, and, together with millions of others, be buried in ocean sediment and turned into limestone. The net result of this long journey is that our molecule of CO_2, originally vented into the atmosphere during a volcanic eruption, would be stored in limestone on the seafloor. This is the process that keeps the CO_2 supplied by volcanism in balance over the long term, maintaining the atmospheric content at a low level—and gradually building up huge amounts of limestone at the Earth's surface. If all the CO_2 bound up in that limestone were to be released, our atmosphere would be more like that of Venus, and the Earth too would be scorchingly hot because of the greenhouse effect.

With an understanding of how atmospheric CO_2 is controlled, it is reasonable to ask whether any of the processes we've discussed can be implicated in starting or ending specific ice ages. The Permo-Carboniferous, as already indicated, may be tied to the lowering of the carbon dioxide content by vascular plants. We know, too, that the amount of volcanic activity, the source of atmospheric CO_2, has varied significantly through geologic time, but to the extent that it's possible to track these changes, there is no evidence that connects any of the known ice ages with reduced volcanism. However, there is an interesting and plausible scenario that has been proposed for the initiation of the Pleistocene Ice Age, one that links it with increased chemical weathering and a consequent decrease in atmospheric CO_2. This particular suggestion—by no means proven, and at this point very much a working hypothesis—also illustrates once again the multiple interconnections among geologic processes, because it invokes the creation of the Himalayan Mountains and uplift of the Tibetan Plateau as its starting point. Formation of the Himalayas is well understood: it is the result of plate tectonics. Three hundred million years ago, India was part of the Gondwanaland supercontinent near the South Pole, and was partly covered with the ice sheets of the Permo-Carboniferous Ice Age. As the large continent broke up, India slowly drifted north and eventually crashed into Asia, with the crumpled and pushed up rocks caught up in the collision forming the Himalayas. The collision began about 50 million years ago and continues today—India is still pushing north against Asia. The timing of mountain building in the Himalayas coincides closely with the beginning of glaciation in the Antarctic 35 million years ago, and uplift continued throughout the gradual cooling of the planet toward the Pleistocene Ice Age.

It's possible that the close agreement in timing between Himalayan mountain building and global cooling toward the Pleistocene Ice Age is pure coincidence. But there are several reasons why the idea is worth serious consideration. One is that high-altitude regions are cold and better able to sustain glaciers than lowlands in any state of the global climate, and in the specific case of the Himalayas and the Tibetan Plateau, the area at high altitude is enormous, almost half the size of the United States. As ice cover increased, so did the positive feedback effect of reflected solar energy. The sheer size and height of the Tibetan Plateau also means that its presence completely altered the global wind pattern, resulting in significant regional changes in climate, although these would have been only indirectly involved in the initiation of glaciation. However, the cornerstone of the proposal is that mountainous regions are sites of intense chemical weathering. Because this process removes CO_2 from the atmosphere, a significant increase in chemical weathering would cause global cooling by reducing the greenhouse effect.

There is no question that chemical weathering is more rapid in the mountains than on flat plains. Anyone who has spent time in mountain ranges, especially geologically young ones such as the Himalayas, or the Rockies or Alps, knows that there are shattered and broken up rocks everywhere. Piled up against every steep cliff is a mound of talus, loose fragments of broken rock that are much more easily attacked by acidic rainwater than is the solid bedrock of flatter regions. Furthermore, mountains are natural rainmakers. When moisture-laden winds are forced upward along mountain fronts, the air cools, water vapor condenses to liquid drops, and it rains. This general principle is accentuated in the Himalayas, because as the vast Tibetan Plateau heats up in the summer sun, the air above it rises, pulling in moist air from the tropical Indian Ocean. The result is the famous Indian monsoons, which drench the mountain front and further accelerate weathering.

All of the available evidence suggests that uplift of the Himalayas has caused chemical weathering in the region to increase greatly. Today, the rivers draining the mountain range all carry much higher amounts of dissolved material than most other world rivers; in spite of the fact that their watersheds cover a relatively small fraction of the Earth's total surface area, they deliver almost a quarter of all the dissolved material flowing into the oceans. The great pile of sediments in the Bay of Bengal, at the mouth of the Ganges, further documents the erosive power of the monsoon rains.

Thus it is likely that uplift and weathering of the Himalayas has affected atmospheric CO_2 and played a part in the gradual cooling of our planet over the past 35 million years. In the most comprehensive computer simulations of climate the influence of greenhouse gases is quite clear: the coldest temperatures and greatest ice cover always appear in trials with the lowest amounts of CO_2. If mountain building *is* implicated in the initiation of the Pleistocene Ice Age through its effect on chemical weathering, then it may have played a role in ice ages of the more distant past, too.

And if greenhouse gases are so important for ice age climate, what does that say about the future? Today, of course, the concern is about increasing CO_2 and global warming, not cooling. Over the past few thousand years, and accelerating over the past few hundred years, clearing land for agriculture, burning forests, and especially burning fossil fuels, has added CO_2 to the atmosphere more quickly than it can be taken up by weathering or photosynthesis or dissolution in the ocean. Its concentration has risen by about one third, and is still increasing rapidly, now almost entirely due to burning of fossil fuel. During the last interglacial period, some 120,000 years ago, CO_2 levels (and temperatures) were similar to those of today. But the greenhouse gas content of the atmosphere did not rise any higher—in fact, it decreased as the Earth began cooling into the glacial interval that culminated only 20,000 years ago. At present, already at a peak concentration, CO_2 is being added to the atmosphere a hundred times faster than it was during any of the natural increases that can be observed in ice cores.

Even critics of global warming don't dispute the greenhouse property of CO_2—it is a well-known fact of physics. It is difficult to comprehend how further additions to the atmosphere at current rates could fail to raise global temperatures and possibly influence the course of the Pleistocene Ice Age. And there is an additional possible source of climate surprise that may arise because of the increasing temperatures: methane. We saw in chapter 8 that release of large amounts of methane has been suggested as a cause for the end of Snowball Earth ice ages and the very warm periods that immediately followed. Today, there are great stores of methane locked up as "hydrates"—icy crystals that contain large amounts of methane in a relatively small volume. The methane in this unusual form of ice comes primarily from bacterial action, and it forms solid layers in the permafrost of cold regions and also in ocean sediments along the margins of continents. The hydrate crystals can only exist over a restricted range of low temperatures and moderate pressures, and decompose easily when conditions change.

As global temperatures increase, heat penetrates slowly into the arctic permafrost and the oceans. Both gradually warm up, and at some point, the hydrates will become unstable and begin to release their trapped methane, which could trigger abruptly increased warming. Even though methane has a short lifetime in the atmosphere, its greenhouse effect would produce a sharp upward temperature spike and could be prolonged if the very large amounts of existing hydrate were to decompose sporadically over a period of time.

There are indications in the geological record that sudden bursts of methane have been released into the atmosphere in the past. The physical evidence includes "pockmarked" sediments in the Arctic and sub-Arctic—areas where detailed mapping of the seafloor shows multiple craters up to a hundred meters across, interpreted to be the result of rapid release of large bubbles of methane gas, probably due to the decomposition of hydrate layers. Destruction of the hydrates most likely resulted from the gradual warming of seawater during the present interglacial period. The chemical composition of some ocean sediments also points to large-scale methane release. Carbon in the methane produced by bacteria has a very distinctive isotopic makeup, and when it is released into seawater that signature is transferred to organisms living in the water, and eventually gets preserved in the sediments. Along the central California coast and elsewhere, recent sediments show series of isotopic "spikes" that appear to be attributable to abrupt injection of methane into seawater—presumably from the decomposition of hydrates. And much farther back in the geologic record, about 55 million years ago, one of the largest recorded abrupt increases in ocean water temperature—7 to 8 degrees Celsius—is accompanied by similar isotopic evidence for methane release. Most scientists have concluded that huge volumes of methane hydrates must have suddenly decomposed, for reasons that are still unclear, and that the methane release was responsible for the sudden temperature increase that followed.

How effectively methane has contributed to the warming of the Earth during the present interglacial period is still a topic of debate. What its role will be in the future is also uncertain. But two things are clear: first, there are very large stocks of this gas stored both on land at high latitudes and along the continental shelves almost everywhere; and second, there is an undeniable correlation in the ice-core data between increased temperature and increased methane in the atmosphere. Even if methane is not the immediate cause, it follows temperature increases very closely and must amplify them.

In spite of the well-documented rise in atmospheric CO_2 and the possibility that large amounts of methane gas will also be released, the consensus view until recently has been that the current warm interglacial period will end soon (in geological terms) and that the Earth is headed toward another glacial episode. This conclusion was based mainly on examination of the past climate record—the peak of the last glaciation was twenty thousand years ago, and over the past million years or so the warm interglacial periods that separate major ice advances have typically lasted only ten or twenty thousand years. Man's additions of CO_2 to the atmosphere may prolong the warm climate of the current interglacial period a bit, but at current rates of usage, our supply of fossil fuels will run out in a few centuries anyway. Elevated levels of carbon dioxide will linger in the atmosphere long after that, but will gradually decrease, reducing the greenhouse effect. Inexorably, the fluctuations in the Earth's orbit will draw us into the next glacial episode, and ice sheets will once again build up from centers in Scandinavia, Canada, and Russia.

Or will they? It is possible that the consensus view is wrong. We have seen how the glacial-interglacial cycles of the past million years have closely followed the 100,000-year timescale of the eccentricity of the Earth's orbit, its tendency to be more or less elliptical. Although exactly why climate tracks eccentricity is not known with certainty, the correlation is clear. And a close look at how the Earth's orbit will change in the future shows that its eccentricity will decrease steadily to almost zero about 30,000 years from now. This is apparent even in James Croll's original graph, reproduced in chapter 5 (figure 11). It is something that has not happened for hundreds of thousands of years. The practical effect is that the variability in the amount of solar radiation received by the Earth will be much less over the next 50,000 years or so than it has been through the past few glacial-interglacial cycles. Coupled with persisting high levels of CO_2, this could push the next glacial advance far into the future. Some computer simulations suggest that under these conditions, significant glaciation will not occur before sixty or seventy thousand years from now, and even then the ice will not be as extensive as it was during the previous few glacial advances. And there is yet another possibility. If CO_2 emissions are not curbed, global warming could completely melt the Greenland glaciers and a substantial part of the Antarctic ice sheet. This would not happen instantaneously; the melting would continue over many human generations. Nevertheless, the consequences for mankind would be serious: the sea level would rise by nearly sixty meters, flooding vast areas of the continents, including most parts of present-day cities like New York and London; weather patterns worldwide would be altered drastically, disrupting agriculture in unpredictable ways; the frequency and intensity of hurricanes would increase because they draw their energy from warm ocean water, which would be far more widespread than currently. Warming would be reinforced by the loss of highly reflective ice and snow, and possibly by the decomposition of unstable methane hydrates. The elevated temperatures coupled with complete loss of continental ice sheets might constitute a threshold-crossing event that would thrust the Earth into a regime from which the glaciers could not quickly recover, even with the return of greater eccentricity and lower CO_2 levels. Only a few hundred years after Louis Agassiz announced his theory of a global ice age, mankind may inadvertently bring the Pleistocene Ice Age to a premature close, ushering in another long period of ice-free existence for our planet.

SUGGESTIONS FOR
FURTHER READING

ICE AGES AND GLACIATION, GENERAL

John C. Crowell, *Pre-Mesozoic Ice Ages: Their Bearing on Understanding the Climate System* (Boulder, CO: Geological Society of America, 1999). This is Memoir 192 of the Geological Society of America. Crowell has spent a distinguished career studying ice ages and here uses his immense expertise to sift through and summarize the disparate evidence for each of the Earth's ancient ice ages and to search for their causes and connections to the climate system.

M. J. Hambrey, *Glacial Environments* (Vancouver: University of British Columbia Press, 1994). A well-illustrated treatment of the effects of glaciers on landscape.

J. Imbrie and K. P. Imbrie, *Ice Ages: Solving the Mystery* (Short Hills, NJ: Enslow, 1979). A well-written account of how ideas about ice ages developed, with insights (by one of the participants) into the work on sediment cores that confirmed the astronomical controls on Pleistocene glaciation.

R. A. Muller and Gordon J. Macdonald, *Ice Ages and Astronomical Causes* (New York: Springer, 2000). A technical and mathematical analysis of the evidence for astronomical control of ice ages.

LOUIS AGASSIZ

Louis Agassiz, *Studies on Glaciers, Preceded by the Discourse of Neuchâtel,* ed. and trans. by Albert V. Carozzi (New York: Hafner, 1967). This is an English translation of Agassiz's famous *Études sur les glaciers,* originally published in 1840. It also includes the text of Agassiz's address to the Natural History Society of Switzerland in Neuchâtel in 1837. The translation includes the magnificent drawings that accompanied the original book.

Edward Lurie, *Louis Agassiz: A Life in Science* (Chicago: University of Chicago Press, 1960). A comprehensive scholarly account of Louis Agassiz's life. However, it focuses mainly on his contributions to biology, with very little discussion devoted to the theory of ice ages.

Jules Marcou, *Life,Letters, and Works of Louis Agassiz* (New York: Macmillan, 1896). A very detailed account of Agassiz's life written by a colleague and personal friend who is perhaps a little biased in his treatment. Although the book is written in English, Marcou reproduced many of Agassiz's letters in their original French.

JAMES CROLL

James Croll, *Climate and Time in Their Geological Relations: A Theory of Secular Changes of the Earth's Climate* (London: Daldy, Isbister, 1875). Croll's masterpiece, in which he brings together his ideas about the Earth's climate.

J.C. Irons, *Autobiographical Sketch of James Croll, with Memoir of his Life and Work* (London: Edward Stanford, 1896). A sympathetic account written by a friend who wished to make the remarkable details of Croll's life known to a wider audience. It includes a listing of all of Croll's publications. It is still the only biography available.

MILUTIN MILANKOVITCH

Milutin Milankovitch, *Cannon of Insolation and the Ice-Age Problem* (Jerusalem: Israel Program for Scientific Translations, 1969). Originally published in 1941 in Belgrade as *Kanon der Erbestrahlund and seine Anwendung auf das Eiseitenproblem,* this was Milankovitch's culminating effort to bring together all of his calculations and ideas about the Earth's climate. Much more mathematically based than James Croll's *Climate and Time,* it is, like that earlier book, a masterpiece.

Milutin Milankovitch, *Milutin Milankovitch 1879–1958* (European Geophysical Society, 1995). This slim volume documenting Milankovitch's life was put together by his son, Vasko, after his father's death. It draws heavily on Milankovitch's autobiography and is well-illustrated with photographs.

THE CHANNELED SCABLANDS

V. R. Baker, ed. *Catastrophic Flooding: The Origin of the Channeled Scabland.* Benchmark papers in Geology, 55 (Stroudsburg, PA: Dowden, Hutchinson & Ross, 1981). This compilation includes the important scientific papers (mostly authored by J. Harlan Bretz) that led to the acceptance of a catastrophic flood origin for the Channeled Scablands, each preceded by a commentary written by the editor.

SNOWBALL EARTH

Paul F. Hoffman and Daniel P. Schrag, "Snowball Earth" (*Scientific American,* January 2000). Hoffman has been the leading proponent of the Snowball Earth hypothesis, and here he and his Harvard colleague Daniel Schrag present their arguments in clear and convincing language.

Gabrielle Walker, *Snowball Earth: The Story of the Great Global Catastrophe That Spawned Life As We Know It* (New York: Crown Publishers, 2003). Walker accompanied Hoffman to some of the important geological localities that provide evidence for the Snowball Earth hypothesis. In this book she gives a lively and very readable account that focuses (positively) on Hofmann's ideas but also touches on those of some of his opponents.

ICE AGES AND EVOLUTION

William H. Calvin, *The Ascent of Mind: Ice Age Climates and the Evolution of Intelligence* (New York: Bantam Books, 1991). Calvin argues that the evolution of human intelligence was stimulated by the ice age climate of Africa. He focuses especially on the implications of fluctuating climate (and the consequent effects on vegetation) for human behavior and brain size.

William H. Calvin, *A Brain for All Seasons: Human Evolution and Abrupt Climate Change* (Chicago: University of Chicago Press, 2002). Calvin updates the arguments of his previous book (above) with new evidence from ice cores for extremely rapid climate fluctuations.

Steven M. Stanley, *Children of the Ice Age: How a Global Catastrophe Allowed Humans to Evolve* (New York: Harmony Books, 1996). Stanley argues that the ice age climate in Africa was a key element in the evolution of humans.

CLIMATE AND HISTORY

Brian Fagan, *The Little Ice Age: How Climate Made History 1300–1850* (New York: Basic Books, 2000). A delightful and fact-packed book that details the chronology of events during the period of the Little Ice Age.

P. D. Jones, A. E. J. Ogilvie, T. D. Davies, and K. R. Briffa, eds., *History and Climate* (New York: Kluwer Academic/Plenum Publishers, 2001). A series of academic papers on many aspects of climate and history, from effects on agriculture to the spread of disease. The contributions are based on papers given at the Second International Climate and History Conference, which took place at the University of East Anglia in the United Kingdom in September 1998.

H. H. Lamb, *Climate, History and the Modern World* (1982; 2d ed., New York: Routledge, 1995). A classic study of the effects of climate on world history.

ABRUPT CLIMATE CHANGE

Richard B. Alley, *The Two-Mile Time Machine: Ice Cores, Abrupt Climate Change, and Our Future* (Princeton: Princeton University Press, 2000). Alley has spent much of his career working on ice cores and in this book provides an enthusiastic insider's view of the harsh working conditions at the Greenland Ice Cap and the excitement and scientific rewards that accrue from unraveling the story contained there. He also reflects on the implications of these records for our future.

Committee on Abrupt Climate Change, National Research Council, *Abrupt Climate Change: Inevitable Surprises* (Washington, D.C.: National Academy Press, 2002). This book, compiled by a committee of the National Research Council (U.S.A.), presents the evidence for abrupt climate change that was available at the time of its publication and makes recommendations about further research necessary to help deal with such changes should they occur in the future.

INDEX

Text:	11/15 Granjon
Display:	Granjon
Compositor:	TechBooks
Printer and Binder:	Edwards Brothers, Inc.